Holt
Mathematics

Chapter 12 Resource Book

ISBN 0-03-078309-7

6   170   09 08

CONTENTS

Holt Mathematics

Holt Mathematics

Date ___________

Dear Family,

In this chapter, your child will learn important problem solving skills to use in working with equations and inequalities. The problem solving skills taught will be applied in consumer education, health issues, physical science, social studies and sports.

When you solve equations with one or more operations, you use inverse operations to isolate the variable.

One-Step Equation

$$n + 7 = 15$$
$$\underline{-7 \quad -7}$$
$$n = 8$$

You need to use another inverse operation.

Two-Step Equation

$$2x + 3 = 23$$
$$\underline{\qquad -3 \quad -3}$$
$$2x \qquad = 20$$
$$\frac{2x}{2} = \frac{20}{2}$$
$$x = 10$$

Some equations have variable terms on both sides of the equal sign.

Solving Equations with Variables on Both Sides

$$6m = 4m + 12$$
$$6m - 4m = 4m - 4m + 12 \qquad \textit{Subtract 4m from both sides.}$$
$$2m = 12 \qquad \textit{Simplify.}$$
$$\frac{2m}{2} = \frac{12}{2} \qquad \textit{Divide both sides by 2.}$$
$$m = 6$$

Your child will have the opportunity to see how two-step algebraic equations are used in a real-life situation.

Marilyn can buy a video game console for \$74.80 and rent a game for \$1.99 per day, or she can rent a console and the same game for \$4.19 per day. How many days would Marilyn have to rent both the video game and the console to pay as much as she would if she had bought the console and rented the game instead?

Let *d* represent the number of days.

$$4.19d = 74.80 + 1.99d$$
$$4.19d - 1.99d = 74.80 + 1.99d - 1.99d \qquad \textit{Subtract 1.99d from both sides.}$$
$$\frac{2.20d}{2.20} = \frac{74.80}{2.20} \qquad \textit{Simplify. Divide both sides by 2.20}$$

$$d = 34$$

Marilyn would need to rent both the video game and the console for 34 days to pay as much as she would have if she had bought the console.

Holt Mathematics

After the work on equations, your child will learn about inequalities. An inequality states that two values are not equal or may not be equal. An inequality uses one of the following symbols:

Symbol	Meaning	Word Phrases
>	Less than	Fewer than, below
<	Greater than	More than, above
≤	Less than or equal to	At most, no more than
≥	Greater than or equal to	At least, no less than

Write an inequality for each situation.

There are at least 25 students in the auditorium.
number of students $\geq$ 25   *"At least" means greater than or equal to.*

There are fewer than 12 people in line.
number of people $<$ 12   *"Fewer than" means less than.*

No more than 150 people can occupy the room.
room capacity $\leq$ 150   *"No more than" means less than or equal to.*

An inequality can be graphed on a number line by drawing a line through all of the values that make the inequality true.

If the inequality is **"greater than"** or **"less than,"** then the number is not included in the graph. This is indicated with an open circle on the number line.

$$y > 2$$

$$-5\ -4\ -3\ -2\ -1\ \ 0\ \ 1\ \ 2\ \ 3\ \ 4\ \ 5$$

If the inequality is **"greater than or equal to"** or **"less than or equal to,"** the number is included in the graph. This is indicated with a closed circle on the number line.

$$y \geq 2$$

$$-5\ -4\ -3\ -2\ -1\ \ 0\ \ 1\ \ 2\ \ 3\ \ 4\ \ 5$$

After your child becomes familiar with the meanings of inequalities and their graphs, he or she will solve inequalities such as $\frac{x}{11} \leq 3$ and $-y + 17 > -10$. Your child will also be asked to graph the solution sets for some of these inequalities.

For additional resources, visit go.hrw.com and enter the key word MS7 Parent.

Holt Mathematics

Practice A
LESSON 12-1
Solving Two-Step Equations

Solve each equation. Cross out each number in the box that matches a solution.

6	−8	−4	−6	−18	−3	2	4	8	3	18

1. $5x + 8 = 23$

2. $-2p - 4 = 2$

3. $6a - 11 = 13$

____

4. $4n + 12 = 4$

5. $9g + 2 = 20$

6. $\dfrac{k}{6} + 8 = 5$

____

7. $\dfrac{s}{3} - 4 = 2$

8. $\dfrac{c}{2} + 5 = 1$

9. $9 + \dfrac{a}{6} = 8$

____

Solve. Check each answer.

10. $3v - 12 = 15$

11. $8 + 5x = -2$

12. $\dfrac{d}{4} - 9 = -3$

____

13. An electrician charges $50 to come to your house. He also charges $25 for each hour he spends at your house. The electrician charges you a total of $125. How many hours did he spend at your house?

____

Holt Mathematics

LESSON 12-1 Practice B
Solving Two-Step Equations

Solve. Check each answer.

1. $7x + 8 = 36$

2. $-3y - 7 = 2$

3. $4a - 13 = 19$

____

____

____

4. $6a - 4 = -2$

5. $5k + 2 = 6$

6. $9m - 14 = -8$

____

____

____

Solve.

7. $\frac{v}{4} - 3 = 5$

8. $\frac{u}{5} + 3 = 1$

9. $6 + \frac{z}{9} = 9$

____

____

____

10. $-7 + \frac{f}{2} = -1$

11. $9 + \frac{w}{4} = -5$

12. $\frac{e}{7} - 3 = -5$

____

____

____

13. $-8 + \frac{d}{5} = 2$

14. $\frac{u}{5} + 3 = 6$

15. $\frac{f}{-3} + 5 = 8$

____

____

____

16. Two years of local Internet service costs $685, including the installation fee of $85. What is the monthly fee?

____

4

Holt Mathematics

Practice C
Solving Two-Step Equations

Solve. Check each answer.

1. $15h - 4 = 6$

2. $25k + 12 = -13$

3. $8m - 14 = -7$

____

____

____

4. $-5m - 6 = 84$

5. $3f + 8 = 28$

6. $5z + 6 = -20$

____

____

____

Solve.

7. $\dfrac{d}{14} - 9 = 5$

8. $\dfrac{a}{35} + 20 = 18$

9. $\dfrac{3}{4} - 9p = -\dfrac{3}{5}$

____

____

____

10. $-2\dfrac{2}{3} + \dfrac{d}{6} = \dfrac{2}{3}$

11. $-\dfrac{1}{4} + \dfrac{f}{8} = \dfrac{1}{2}$

12. $-\dfrac{2}{3} = 6 + \dfrac{v}{-3}$

____

____

____

Translate each equation into words, and then solve the equation.

13. $7 + \dfrac{n}{2} = 10$

14. $4w - 9 = -5$

____

____

____

____

15. A taxi charges $1.50 plus a fee of $0.60 for each mile traveled. If a ride costs $5.40, how many miles was the ride?

____

Holt Mathematics

<table><tr><td>LESSON
12-1</td><td>

Reteach
Solving Two-Step Equations
</td></tr></table>

You can solve two-step equations by undoing one operation at a time. First undo any addition or subtraction, then undo any multiplication or division.

Complete the steps to solve each equation.

1. $7x + 3 = 31$

$7x + 3 - \underline{\quad} = 31 - \underline{\quad}$ ⟵ Subtract ___ from both sides to undo addition.

$7x = 28$

$\dfrac{7x}{\quad} = \dfrac{28}{\quad}$ ⟵——— Divide both sides by ___ to undo multiplication.

$x = 4$

Check

$7x + 3 = 31$

$7(\underline{\quad}) + 3 \stackrel{?}{=} 31$ ⟵——— Substitute ___ for x.

$\underline{\quad} + 3 \stackrel{?}{=} 31$

$31 \stackrel{?}{=} 31$ ✔ ⟵——— 4 is a solution.

2. $\dfrac{n}{6} - 8 = 4$

$\dfrac{n}{6} - 8 + \underline{\ } = 4 + \underline{\ }$

$\dfrac{n}{6} = 12$

$6 \cdot \dfrac{n}{6} = \underline{\ } \cdot 12$

$n = \underline{\ }$

3. $8a - 5 = 11$

$8a - 5 + \underline{\ } = 11 + \underline{\ }$

$8a = \underline{\ }$

$\dfrac{8a}{\quad} = \dfrac{16}{\quad}$

$a = \underline{\ }$

4. $9 + \dfrac{w}{2} = 12$

$9 - \underline{\ } + \dfrac{w}{2} = 12 - \underline{\ }$

$\dfrac{w}{2} = \underline{\ }$

$2 \cdot \dfrac{w}{2} = \underline{\ } \cdot 3$

$w = \underline{\ }$

Solve.

5. $4n + 11 = 27$

6. $\dfrac{z}{7} - 6 = 3$

7. $3 - 2k = -7$

____

____

____

6

Holt Mathematics

LESSON 12-1 Challenge
Consecutive Number Search

Consecutive numbers are numbers that come one after the other.

For example, 3 and 4 are consecutive whole numbers.

 6 and 8 are consecutive even numbers.

 7 and 9 are consecutive odd numbers.

If n = a number, then $n + 1$ is the next consecutive whole number after n.

If n = an even number, then $n + 2$ is the next consecutive even number after n.

If n = an odd number, then $n + 2$ is the next consecutive odd number after n.

Solve. (*Hint:* Let n = the first number. Then write an equation for each sum.)

1. The sum of two consecutive numbers is 73. What are the numbers?

____

2. The sum of two consecutive numbers is 131. What are the numbers?

____

3. The sum of two consecutive even numbers is 70. What are the numbers?

____

4. The sum of two consecutive even numbers is 158. What are the numbers?

____

5. The sum of two consecutive odd numbers is 108. What are the numbers?

____

6. The sum of two consecutive odd numbers is 180. What are the numbers?

____

Holt Mathematics

Problem Solving

LESSON 12-1

Solving Two-Step Equations

Write the correct answer.

1. Last week, Carlie had several rice cakes and 3 granola bars as snacks. The snacks contained a total of 800 calories. If each granola bar had 120 calories and each rice cake had 40 calories, how many rice cakes did she have?

2. Jo eats 2,200 calories per day. She eats 450 calories at breakfast and twice as many at lunch. If she eats three meals with no snacks, which meal will contain the most calories?

3. Erika is following a 2,200 calorie-per-day diet. She eats the recommended 9 servings of breads and cereals, averaging 120 calories per serving. She also eats 5 servings of vegetables. If the rest of her daily intake is 870 calories, what is the average number of calories in each serving of vegetables?

4. Brandon follows a 2,800 calorie-per-day diet. He has 11 servings of breads and cereals, which average 140 calories each. Yesterday, he had a combined 9 servings of fruits and vegetables, averaging 60 calories each. How many 180-calorie servings of meat and milk did he have to complete his diet?

Choose the letter for the best answer.

The table shows calories burned by a person performing different activities.

Calories Used in Activities

Activity	Calories (per min)
Basketball	7.5
Cycling (10 mi/h)	5.5
Jogging	9.3
Swimming	7.8

5. Kamisha swims for 0.25 hour. How many calories does she burn?

   A 30 calories   C 1.95 calories

   B 195 calories   D 117 calories

6. Stu jogs at a rate of 5 mi/h. How far must he jog to burn 418.5 calories?

   F 9 mi   H 3.75 mi

   G 4.65 mi   J 45 mi

7. Terry rides her bike for 40 minutes and plays basketball for an hour. How many calories does she burn?

   A 67 calories   C 670 calories

   B 560 calories   D 1,300 calories

8. How many hours would you have to ride your bike at 10 mi/h to burn 550 calories?

   F 1.67 hr   H 1.0 hr

   G 1.5 hr   J 0.75 hr

Holt Mathematics

LESSON 12-1 Reading Strategies
Follow a Procedure

To solve two-step equations, follow these steps.

To Solve Two-Step Equations

$3n + 5 = 23$

Step 1: Get the variable term by itself. Use the inverse operation.

$3n + 5 - 5 = 23 - 5$
$3n = 18$

Subtract 5 from both sides.

Step 2: Get the variable by itself. Use the inverse operation.

$$\frac{3n}{3} = \frac{18}{3}$$

Divide both sides by 3.

Step 3: Compute and simplify the solution.

$n = 6$

Answer the following questions.

1. What is the first step in solving a two-step equation?

____

____

2. Which term in the equation above does not contain a variable?

____

3. What operation was performed to remove that term?

____

____

4. What is the second step in solving a two-step equation?

____

5. Which term in the equation contains a variable?

____

6. What operation was performed to get the *n* by itself?

____

7. What is the third step in a two-step equation?

____

Holt Mathematics

Puzzles, Twisters & Teasers

LESSON 12-1

Sun, Sand, and Snakes!

Solve each equation. Then use the letters of the variables to answer the riddle.

1. $\dfrac{s}{-6} - 21 = -28$ s = _____

2. $-4n - 7 = 17$ n = _____

3. $6a + 4 = 10$ a = _____

4. $\dfrac{i}{3} - 10 = -6$ i = _____

5. $7t - 8 = 27$ t = _____

6. $\dfrac{c}{8} + 25 = 20$ c = _____

7. $-9e - 15 = 93$ e = _____

8. $-3o - 8 = -35$ o = _____

9. $5r + 13 = 15$ r = _____

10. $6m + 4 = 22$ m = _____

What kind of shoes do snakes wear to the beach?

W ___ ___ ___   ___ ___ ___ ___ ___ ___ ___ ___ ___ ___

    1   5   −12 $\dfrac{2}{5}$ 3   9   −40   −40   1   42   12   −6   42

Holt Mathematics

LESSON 12-2 Practice A
Solving Multi-Step Equations

Solve. Choose the letter for the best answer.

1. $12n - 7 - 3n = 11$

A $n = 1$ **C** $n = 9$
B $n = 2$ **D** $n = 18$

2. $3a + 2 + 4a = 23$

F $a = 3$ **H** $a = 7$
G $a = 5$ **J** $a = 9$

3. $4(y + 5) + 3 = 35$

A $y = 3$ **C** $y = 20$
B $y = 4.5$ **D** $y = 27$

4. $2(h - 6) + 20 = -4$

F $h = 4$ **H** $h = -6$
G $h = 1$ **J** $h = -18$

Solve.

5. $x + 6 + 2x = 15$

6. $10b + 9 - 3b = 2$

7. $5n - 2 + 3n = 5$

____

8. $6(w - 8) + 16 = 4$

9. $2(z + 5) + 4 = -12$

10. $1.5(b + 6) + 9 = 24$

____

11. Jose ran twice as many kilometers as Karen. Adding
8 to the number of kilometers Jose ran and dividing by
4 gives the number of kilometers Maria ran. Maria ran
3 kilometers. How many kilometers did Karen run?

____

Holt Mathematics

LESSON 12-2 | Practice B
Solving Multi-Step Equations

Solve.

1. $15x - 8 - 3x = 16$

2. $5n + 3 + 4n = 30$

3. $h - 6 + 7h = 42$

4. $-3g + 6 + 2g = 15$

5. $-2b + 7 - 3b = 2$

6. $5y + 1 + 3y = -15$

7. $4k - 14 + 3k = 21$

8. $9m + 10 - 14m = -5$

9. $-2d + 18 - 4d = 60$

10. $3(n + 5) + 2 = 26$

11. $4 - 2(v - 6) = -8$

12. $1.4 - 1.6(t + 6) = 4.6$

13. $2.4(m - 3) + 3.8 = -8.2$

14. $6 = 8(s - \frac{3}{4}) - 20$

15. $5(c + \frac{4}{5}) + 6 = 50$

16. Joel has twice as many CDs as Mariella has. Subtracting 7 from the number of CDs Joel has and dividing by 3 equals the number of CDs Blake has. If Blake has 25 CDs, how many CDs does Mariella have?

____

Holt Mathematics

LESSON 12-2 Practice C
Solving Multi-Step Equations

Solve.

1. $-31x - 18 + 3x = 38$

2. $15n + 16 + 4n = 130$

3. $h - 48 + 17h = 42$

4. $-6g + 7g - 7 = 9$

5. $16 - 12a - 13a = -9$

6. $8y + 43 - 7y = -15$

7. $4(c - 3) - 6 = 10$

8. $3.8(s - 6) + 4.6 = -22$

9. $18(m + \frac{4}{9}) - 16 = 118$

10. $\dfrac{17u - 24}{18} = \dfrac{5}{9}$

11. $\dfrac{0.4n - 1}{7} = 5$

12. $\dfrac{18 - 4t}{0.5} = 20$

13. $\dfrac{1.4m - 4.5}{1.5} = 8.2$

14. $\dfrac{\frac{1}{3}s - 2}{5} = -2$

15. $\dfrac{5.6 - 2a + 1.4}{1.2} = 4.5$

16. There are 5 times as many tomato plants in the garden as there are cucumber plants. The number of pepper plants is 7 more than the sum of the number of tomato and cucumber plants. There are 19 pepper plants in the garden. How many cucumber plants are there? How many tomato plants are there?

Holt Mathematics

LESSON 12-2 **Reteach**
Solving Multi-Step Equations

Some equations require more than two steps to solve. To make
these equations easier to solve, first combine any like terms. Also,
you may need to use the Distributive Property to complete
multiplication that involves parentheses.

Complete the steps to solve each equation.

1. $5a - 7 + 3a = 17$

    _____ $- 7 = 17$ ⟵——————— Combine like terms: $5a + 3a =$ _____.

    $8a - 7 +$ _____ $= 17 +$ _____ ⟵—— Add _____ to both sides of the equation.

    $8a =$ _____

    $\dfrac{8a}{} = \dfrac{24}{}$ ⟵——————Divide both sides of the equation by _____.

    $a =$ _____

2. $4(x + 6) - 15 = 21$

    _____ $(x) +$ _____ $(6) - 15 = 21$ ⟵—— Multiply each term in the parentheses

    by _____.

    $4x + 24 - 15 = 21$

    $4x +$ _____ $= 21$ ⟵—— Combine like terms: $24 - 15 =$ _____.

    $4x + 9 -$ _____ $= 21 -$ _____ Subtract from both sides of the equation

    by _____.

    $4x =$ _____

    $\dfrac{4x}{4} = \dfrac{12}{4}$ ⟵——Divide both sides of the equation

    by _____.

    $x =$ _____

Solve.

3. $9b + 6 - 5b = 18$ 4. $2y - 7 + y = 8$ 5. $-5c - 12 - 5c = 8$

_______   _______   _______

6. $2(r - 3) + 6 = 14$ 7. $8 + 5(s - 1) = 18$ 8. $6(x + 3) + 38 = 2$

_______   _______   _______

     14   **Holt Mathematics**

<table>
<tr><td>LESSON
12-2</td><td></td></tr>
</table>

Challenge
Solve the System

When equations have two variables, you can sometimes find the point where two equations have the same solution—that is, where one value of x and one value of y solve both equations. This is called "solving a system of equations."

To solve a system of equations, you can substitute from one equation into the other.

Example: Solve $x + y = 6$ and $x - y = 2$.

Step 1: Solve one equation for y. $x + y = 6$
$$y = 6 - x$$

Step 2: Substitute that value of y into the second equation.
Solve for x. $x - y = 2$
$$x - (6 - x) = 2$$
$$x - 6 + x = 2$$
$$2x - 6 = 2$$
$$2x = 8$$
$$x = 4$$

Step 3: Substitute the solution for x from Step 2 into the other equation. $x + y = 6$
$$4 + y = 6$$
$$y = 2$$

Step 4: Write the solution as an ordered pair (x, y).
Solution: $(4, 2)$

Solve each system of equations.

1. $y = 3x$
   $x + y = 20$

2. $m = 4n$
   $m + n = 10$

3. $d = c - 2$
   $c + d = 12$

4. $v - w = 2$
   $3w + v = 14$

____

5. $j = k + 3$
   $j + k = 7$

6. $b = a - 9$
   $a + b = 3$

7. $r = 2 - s$
   $2s + r = 9$

8. $e + f = 2$
   $3e + f = 8$

____

Holt Mathematics

LESSON 12-2 **Problem Solving**
Solving Multi-Step Equations

Write the correct answer.

To convert a temperature from degrees Fahrenheit to degrees
Celsius, you can use the formula $(°F - 32)0.56 = °C$.

1. The record high temperature in North Carolina is 110°F. What is the record high in degrees Celsius?

2. The record low temperature in Florida is −2°F. What is the record low in degrees Celsius?

3. The record high temperature in the United States is 134°F. This was recorded in Greenland Ranch, California, on July 10, 1913. What is that temperature in degrees Celsius?

4. The record high in Texas is 120°F. The record low in Texas is −23°F. In degrees Celsius, what is the range between the record high and low temperatures in Texas?

5. When the temperature is 4°C, you need to wear a heavy coat. Write 4°C as degrees Fahrenheit.

6. When the temperature is 28°C, you might want to go to the beach. Write 28°C as degrees Fahrenheit.

Choose the letter for the best answer.

7. Faith spent $78 at Fashion Warehouse. She bought 2 shirts that each cost $17.50 and a pair of shoes. How much did she spend for the shoes?

   A $34.00   C $60.50

   B $43.00   D $113.00

8. Three friends each pay $4.15 to buy a pizza. A basic pizza costs $9.45. Additional toppings cost $1 each. How many toppings were on the pizza?

   F 2 toppings   H 4 toppings

   G 3 toppings   J 5 toppings

9. Todd buys 3 CDs at $16.99 each and a DVD that costs $24.99. He pays with a $100 bill. How much change does he receive?

   A $24.04   C $49.03

   B $8.04   D $67.96

10. Marina bought 4 books. José bought half as many books as Ben bought. Together, the 3 friends bought 13 books. How many books did Ben buy?

   F 9 books   H 3 books

   G 6 books   J 2 books

Holt Mathematics

LESSON 12-2 **Reading Strategies**
Analyze Information

Look at the terms in this equation.

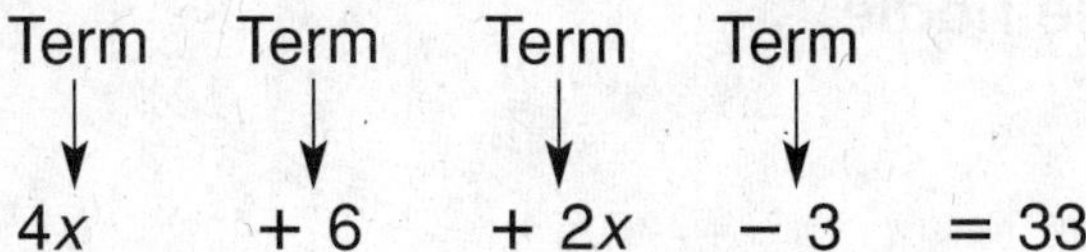

$$4x \quad + 6 \quad + 2x \quad - 3 \quad = 33$$

Answer each question.

1. How many terms does this equation have?

2. How many terms have variables?

3. How many terms do not have variables?

You can rearrange the terms in an equation with like terms together. Terms are "like" if they both have the same variable or do not have a variable.

1st arrangement: $4x \quad + 6 \quad + 2x \quad - 3 \quad = 33$
2nd arrangement: $4x \quad + 2x \quad + 6 \quad - 3 \quad = 33$

Look at the equation to answer each question.

4. Write the two terms that have x as a variable.

5. What is the sum of these two terms?

6. Write the two terms that do not have variables.

7. Find the difference of these two terms to combine them.

8. Rewrite the equation with like terms combined.

Holt Mathematics

Puzzles, Twisters & Teasers
Just a Bunch of Hot Air

Decide whether or not each equation is correct. Circle the letter above your answer. Use the letters to solve the riddle.

1. $b + 18 + 3b = 74$ $b = 29$

   U   **A**

   correct   incorrect

2. $-3 + 4p - 2p = 1$ $p = 4$

   K   **C**

   correct   incorrect

3. $2 - 5(x + 2) = -18$ $x = -4$

   N   **R**

   correct   incorrect

4. $6(n - 7) + 40 = 4$ $n = 1$

   O   **Q**

   correct   incorrect

5. $5n + 7 - 3n = 19$ $n = 3$

   L   **S**

   correct   incorrect

6. $10x - 3 - 2x = 4$ $x = 7$

   G   **S**

   correct   incorrect

7. $2.5(a + 4) - 3.2 = 14.3$ $a = 3$

   W   **P**

   correct   incorrect

8. $10x - 11 - 4x = 43$ $x = 9$

   I   **M**

   correct   incorrect

9. $53 = 12(m - \frac{3}{4}) - 16$ $m = 5$

   J   **N**

   correct   incorrect

10. $7n - 1 - 2n = 14$ $n = 3$

    D   **H**

    correct   incorrect

What is the angriest kind of wind?

____   ____ ____ ____ ____ ____ ____ ____ ____ ____ ____

Holt Mathematics

Practice A

LESSON 12-3

Solving Equations with Variables on Both Sides

Group the terms with the variables on one side of the equal sign and simplify. Do not solve.

1. $9p = 6p + 21$

2. $-5t - 14 = 2t$

3. $2k = 18 - 4k$

____

4. $-6u + 48 = 6u$

5. $7m - 25 = 2m$

6. $\frac{5}{9}a = 6 + \frac{4}{9}a$

____

Solve. Match each equation with its solution.

7. $8n = 6n - 14$ **A.** $n = 8$

8. $7n + 8 = 11n$ **B.** $n = 9$

9. $2n + 8 = 3n$ **C.** $n = -7$

10. $\frac{3}{4}n + 2 = \frac{1}{4}n + 1$ **D.** $n = 2$

11. $9 + 3n = 4n$ **E.** $n = -8$

12. $\frac{5}{8}n + 5 = \frac{7}{8}n + 7$ **F.** $n = -2$

13. Members of the Campus Roller Rink pay a yearly membership fee of $50 plus $5 each time they go skating. Nonmembers pay $10 each time they go skating. How many times would both a member and a nonmember have to go skating in a year in order to pay the same amount? _______________

Holt Mathematics

Practice B
Solving Equations with Variables on Both Sides

Group the terms with the variables on one side of the equal sign and simplify.

1. $10t = 6t + 24$

2. $-6x - 32 = 2x$

3. $j = 20 - 4j$

4. $-5d + 40 = 5d$

5. $9m - 28 = 2m$

6. $\frac{8}{9}x = 8 + \frac{4}{9}x$

Solve.

7. $8k = 6k - 26$

8. $32 - 5v = 3v + 8$

9. $-12y - 10 = -6y + 14$

10. $\frac{5}{8}a + 6 = \frac{3}{4}a$

11. $\frac{1}{4}n + 10 = \frac{2}{3}n$

12. $20 - \frac{1}{5}d = \frac{3}{10}d + 16$

13. Members of the Lake Shawnee Club pay \$40 per summer season plus \$7.50 each time they rent a boat. Nonmembers pay \$12.50 each time they rent a boat. How many times would both a member and a nonmember have to rent a boat in order to pay the same amount? ______________

Holt Mathematics

 # Practice C
 ## *Solving Equations with Variables on Both Sides*

Group the terms with variables on one side of the equal sign and simplify.

1. $15t = 17 + 8t - 24$

2. $6a - 42 = 2a + a$

3. $6k - k = 27 - 4k$

____

4. $-5f + 40 = 5f + 30$

5. $\frac{9}{10}m - 28 = \frac{7}{10}m + 12$

6. $\frac{8}{15}x = 8 + \frac{2}{15}x - \frac{4}{15}x$

____

Solve. Check each answer.

7. $18r - 2 - 12r$
$= -6r - 26$

8. $10 + 8c + 15$
$= -2c - 40$

9. $-3w + 6$
$= 8w + 9 - 2w$

____

10. $0.8a - 6 = -0.6a + 22$

11. $\frac{1}{2}n - 1\frac{5}{12} = \frac{2}{3}n - 2\frac{1}{6}$

12. $\frac{13}{15} - \frac{d}{5} = \frac{5}{6} + \frac{d}{3}$

____

13. A golf course charges nonmembers $48 per round plus a golf cart rental fee of $24 per round. A member pays a monthly fee of $384 plus a golf cart rental fee of $8 per round. How many rounds would both a member and a nonmember have to play in order to pay the same amount? _______________

Holt Mathematics

Reteach
Solving Equations with Variables on Both Sides

Some equations have like terms that are on opposite sides of the equal sign. To solve these equations, first identify the like term with the lesser coefficient.

$2x = 25 - 3x$ ⟵ $2x$ and $-3x$ are like terms.

Since $-3 < 2$, $-3x$ is the term with the lesser coefficient. Move $-3x$ to the opposite side of the equation by adding $3x$ to each side of the equation.

$$2x = 25 - 3x$$
$$2x + \mathbf{3x} = 25 - 3x + \mathbf{3x}$$
$$5x = 25$$
$$\frac{5x}{5} = \frac{25}{5}$$
$$x = 5$$

Complete the steps to solve the equation.

1. $5x + 16 = 9x$

The like terms are _____ and _____.
The like term with the lesser coefficient is _____.

$$5x + 16 \quad = 9x$$
$$5x - \underline{\quad} + 16 = 9x - \underline{\quad}$$
$$16 = \underline{\quad}$$
$$\frac{16}{\quad} = \frac{4x}{\quad}$$
$$\underline{\quad} = x$$

2. $3a + 1 = 15 - 4a$

The like terms are _____ and _____.
The like term with the lesser coefficient is _____.

$$3a + 1 = 15 - 4a$$
$$3a + \underline{\quad} + 1 = 15 - 4a + \underline{\quad}$$
$$\underline{\quad} + 1 = 15$$
$$7a + 1 - \underline{\quad} = 15 - \underline{\quad}$$
$$7a = \underline{\quad}$$
$$\frac{7a}{\quad} = \frac{14}{\quad}$$
$$a = \underline{\quad}$$

Solve.

3. $8d = 7d + 13$

4. $-4p - 16 = 4p$

5. $3n = 25 - 2n$

6. $9c - 6 = 2c + 15$

7. $16 - 6k = 20 - 2k$

8. $2h + 8 = 2 - h$

Holt Mathematics

<table><tr><td>LESSON
12-3</td><td>

Challenge
Use the Clues

</td></tr></table>

You can write equations using clues to solve problems.

Example: The sum of two numbers is 36. The difference of the two numbers is 8. Find the numbers.

Let x = one number.

Let y = the other number.

The sum of the two numbers is 36.	$x + y = 36$
The difference of the two numbers is 8.	$x - y = 8$
Solve the system of equations.	$x + y = 36$
	$x - y = 8$

The solution is (22, 14), so the two numbers are 22 and 14.

Solve.

1. The sum of two numbers is 33. The difference of the two numbers is 5. Find the numbers.

   ____

2. The sum of two numbers is 85. The difference of the two numbers is 13. Find the numbers.

   ____

3. One number is 18 more than another number. The sum of the two numbers is 60. Find the numbers.

   ____

4. The sum of two numbers is 99. The difference of the two numbers is 9. Find the numbers.

   ____

5. The sum of two numbers is 42. One number is twice the other number. Find the numbers.

   ____

6. The sum of two numbers is 43. One number is equal to 5 less than 3 times the other number. Find the numbers.

   ____

Holt Mathematics

Problem Solving

LESSON 12-3

Solving Equations with Variables on Both Sides

Write the correct answer.

1. Five added to twice Erik's age is the same as 3 times his age minus 2. How old is Erik?

2. Three times the perimeter of a triangle is the same as 75 decreased by twice the perimeter. What is the perimeter of the triangle?

3. The area of a pentagon increased by 27 is the same as four times the area of the pentagon, minus 15. What is the area of the pentagon?

4. To repair body damage on a car, AutoBody charges $125, plus $18 per hour. CarCare charges $200, plus $12 per hour. Determine the number of hours for which the two body shops will cost the same.

Choose the letter for the best answer.

5. Sandy and Suzanne are planting flower pots around the school building. Sandy has planted 33 pots and is planting at the rate of 10 pots per hour. Suzanne has planted 25 pots and is planting at the rate of 14 pots per hour. In how many hours will they have planted the same number of flower pots?

   A 3 hr   **C** 2 hr

   B 2.5 hr   **D** 1 hr

6. The length of the sides of a square measure $2x - 5$. The length of a rectangle measures $2x$, and the width measures $x + 2$. For what value of x is the perimeter of the square the same as the perimeter of the rectangle?

   F $x = 2$ **H** $x = 10$

   G $x = 7$ **J** $x = 12$

7. Louisa used Downtown Taxi, which charges $2 for the first mile and $1.10 for each additional mile. Pietro used Uptown Cab, which charges $5 for the first mile and $0.95 for each additional mile. They paid the same amount and traveled the same distance. How far did they travel?

   A 25 mi   **C** 20 mi

   B 21 mi   **D** 15 mi

8. Toni bought some beach towels on sale for $8 each. Theo bought the same number of beach towels at the full price of $12. Toni's total was $24 less than Theo's total. How many beach towels did they each buy?

   F 6 towels   **H** 9 towels

   G 8 towels   **J** 12 towels

Holt Mathematics

<table><tr><td>**LESSON**
12-3</td><td># Reading Strategies
Follow a Procedure</td></tr></table>

To solve equations with variables on both sides, follow these steps.

$$8x - 3 = 2x + 9$$

Step 1: Get all variable terms on one side of the equation.

$$8x - \mathbf{2x} - 3 = 2x - \mathbf{2x} + 9$$
$$6x - 3 = 9$$

Subtract $2x$ from both sides.

Step 2: Isolate the variable term. Use the opposite operation.

$$6x - 3 + \mathbf{3} = 9 + \mathbf{3}$$
$$6x = 12$$

Add 3 to both sides.

Step 3: Use the inverse operation and calculate.

$$\frac{6x}{6} = \frac{12}{6}$$
$$x = 2$$

Divide both sides by 6.

Answer the following questions.

1. What is the first step to solve an equation with variables on both sides?

____

____

2. What operation was performed to get the variables on one side?

____

3. Write the equation that results after the first step is complete.

____

4. What is the second step to solve the equation?

____

____

5. What operation was performed to get 3 on the opposite side?

____

6. Write the equation the way it looks after the second step is complete.

____

7. What operation was performed last?

____

Holt Mathematics

Puzzles, Twisters & Teasers
LESSON 12-3 *All Aboard!*

Decide whether or not each solution is correct. Circle the letter above your answer. Then use the letters to solve the riddle.

1. $6m = 4m + 12$ $m = 6$

    W   **A**
  correct   incorrect

2. $5n = 4n + 32$ $n = 4$

    M   **I**
  correct   incorrect

3. $4y = 2y + 40$ $y = 10$

    V   **D**
  correct   incorrect

4. $5n = 3n + 26$ $n = 13$

    E   **Q**
  correct   incorrect

5. $8 + 6a = -2a + 24$ $a = 3$

    J   **O**
  correct   incorrect

6. $19 + 7n = -2n + 37$ $n = 2$

    N   **W**
  correct   incorrect

7. $12h = 9h + 84$ $h = 12$

    P   **T**
  correct   incorrect

8. $\frac{5}{9}x + 83 = \frac{4}{9}x + 92$ $x = 81$

    W   **C**
  correct   incorrect

9. $9t = 4t + 120$ $t = 7$

    G   **A**
  correct   incorrect

10. $4.19d = 74.8 + 1.99d$ $d = 34$

    I   **B**
  correct   incorrect

How do you define a "twip"?

A ____ ____ ____ ____ you take ____ ____ a ____ ____ ____ ____ n

Holt Mathematics

Name _______________________________________ Date ___________ Class ___________

 ## Practice A
Inequalities

Choose an inequality for each situation.

$$x > 10 \qquad x \geq 10 \qquad x < 10 \qquad x \leq 10$$

1. The temperature today will be at least 10°F.   _____________

2. The temperature tomorrow will be no more than 10°F.   _____________

3. Yesterday, there was less than 10 inches of snow.   _____________

4. Last Wednesday, there was more than 10 inches of snow.   _____________

Complete the graph for each inequality.

5. $a > 3$

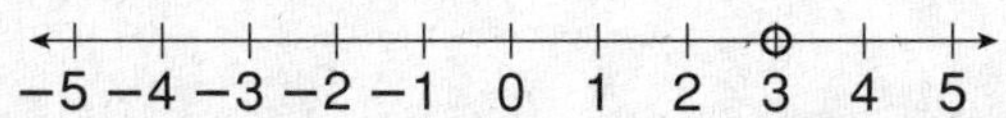

6. $r \leq -2$

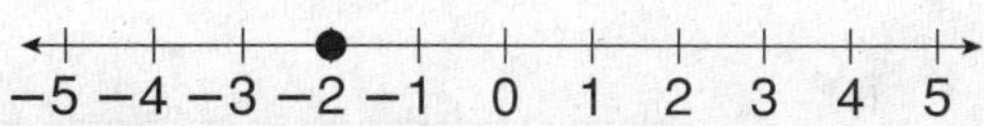

Graph each inequality.

7. $w \geq 0$

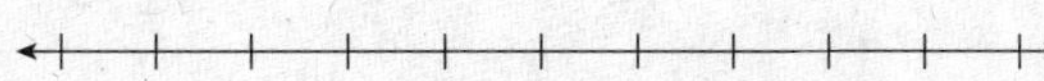

8. $b \leq -4$

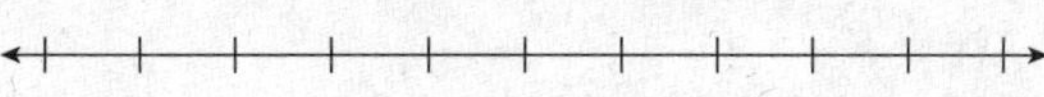

9. $j > 5$

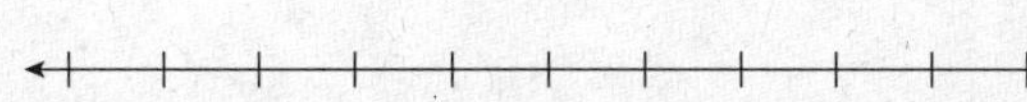

Complete the graph for each compound inequality.

10. $t > 2$ or $t < -1$

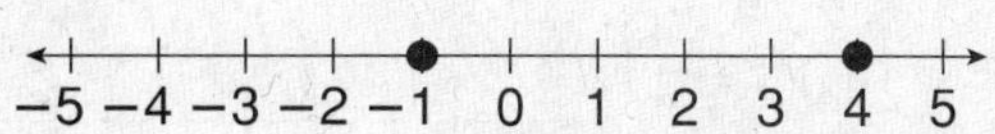

11. $-1 \leq f \leq 4$

Graph each compound inequality.

12. $y \geq 1$ or $y < -3$

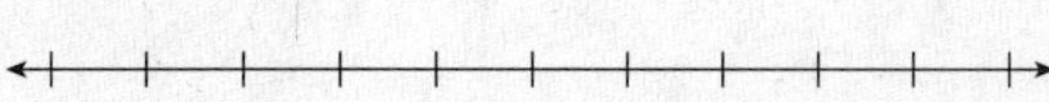

13. $-4 < p < 2$

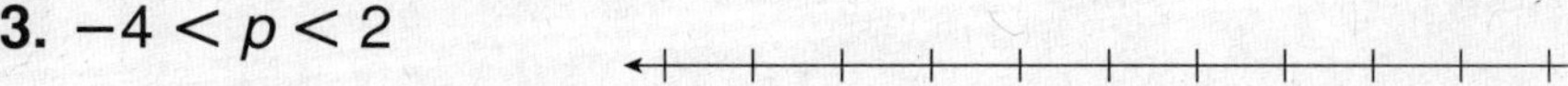

Holt Mathematics

LESSON 12-4 **Practice B**
Inequalities

Write an inequality for each situation.

1. The temperature today will be at most 50°F. _____________

2. The temperature tomorrow will be above 70°F. _____________

3. Yesterday, there was less than 2 inches of rain. _____________

4. Last Monday, there was at least 3 inches of rain. _____________

Graph each inequality.

5. $t \le -2$

6. $j > -5$

7. $y \le 0$

8. $b < \dfrac{1}{2}$

Graph each compound inequality.

9. $f > 3$ or $f < -2$

10. $-4 \le w \le 4$

11. $b < 0$ or $b \ge 5$

12. $y \ge 3$ or $y \le -1$

13. $-4 < m < -2$

Holt Mathematics

LESSON 12-4 Practice C
Inequalities

Graph each inequality or compound inequality.

1. $x > 14$

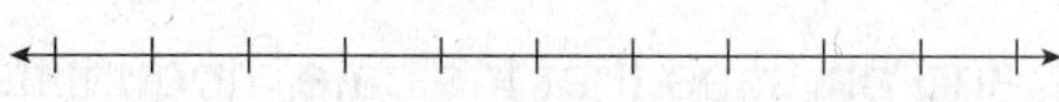

2. $t \le -22$

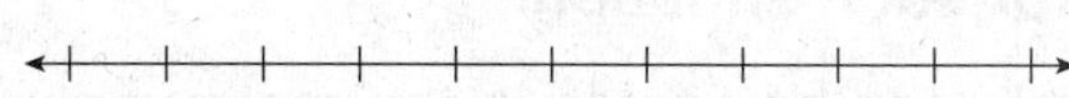

3. $j > -35$

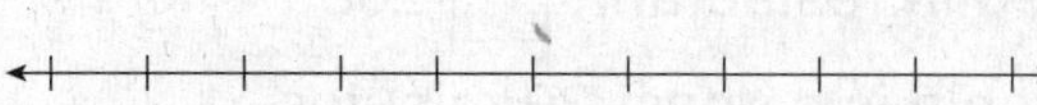

4. $f > 12$ or $f < -12$

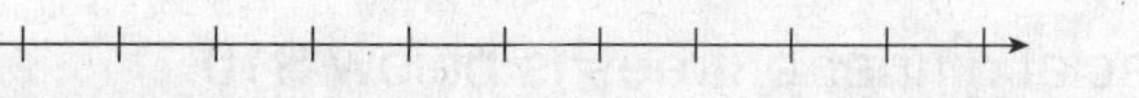

5. $-14 \le w \le 2$

6. $b < 0$ or $b \ge 15$

Write each statement using inequality symbols.

7. The number x is between -9 and -19.

____

8. The number y is at most 5.

____

9. The number g is greater than or equal to 4 or less than -2.

____

10. The number p is less than or equal to -8.

____

Write an inequality shown by each graph.

11.

____

12.

____

13.

____

14.

____

Holt Mathematics

Reteach
Inequalities

An equation is a statement that says two quantities are equal. An **inequality** is a statement that says two quantities are **not** equal.

The chart shows symbols and phrases that indicate inequalities.

<	>	≤	≥
Less than	Greater than	Less than or equal to	Greater than or equal to
Fewer than	More than	At most	At least
Below	Above	No more than	No less than

Complete the inequality for each situation.

1. No more than 200 people can be seated in the restaurant.

    number of people seated in restaurant ☐ 200

2. The waiting time for a table is at least 20 minutes.

    waiting time ☐ 20 minutes

3. The price of all special dinner entrees is below $10.

    special dinner entrees ☐ $10

4. The Yoshida family spent more than $40 for dinner.

    Yoshida family spent ☐ $40

An inequality can be shown on a graph.

The graph shows:	Inequality	Graph
All numbers greater than 3	$x > 3$	−5 −4 −3 −2 −1 0 1 2 3 4 5 The *open circle* at 3 shows that the value 3 is **not** included in the graph.
All numbers greater than or equal to 3	$x \geq 3$	−5 −4 −3 −2 −1 0 1 2 3 4 5 The *closed circle* at 3 shows that the value 3 **is** included in the graph.
All numbers less than 3	$x < 3$	−5 −4 −3 −2 −1 0 1 2 3 4 5
All numbers less than or equal to 3	$x \leq 3$	−5 −4 −3 −2 −1 0 1 2 3 4 5

Holt Mathematics

LESSON 12-4

Reteach
Inequalities (continued)

Graph each inequality.

5. $x > -4$

- Draw an open circle at −4.
- Read $x > -4$ as "x is greater than −4."
- Draw an arrow to the right of −4.

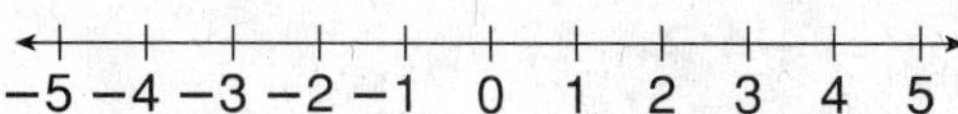

6. $x \leq 1$

- Draw a closed circle at 1.
- Read $x \leq 1$ as "x is less than or equal to 1."
- Draw an arrow to the left of 1.

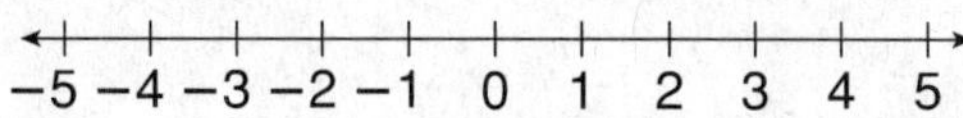

7. $a > -1$

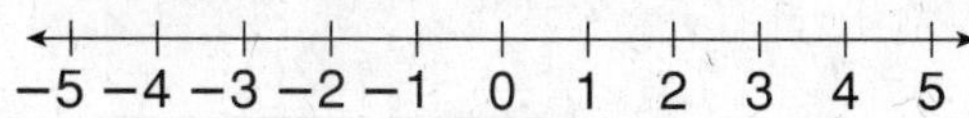

8. $y \leq 3$

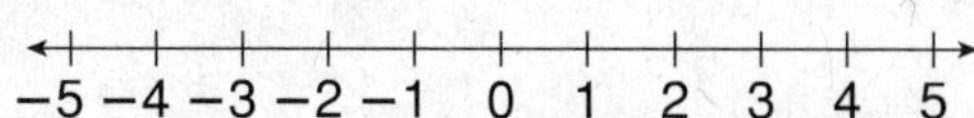

A **compound inequality** is a combination of two inequalities.

The graph shows:	Inequality	Graph
All numbers from −2 to 2	$-2 \leq x \leq 2$	
All numbers greater than 2 *or* less than −2	$x > 2$ or $x < -2$	

Graph each compound inequality.

9. $0 < x \leq 3$

- Draw an open circle at 0 and a closed circle at 3.
- Shade the line between 0 and 3.

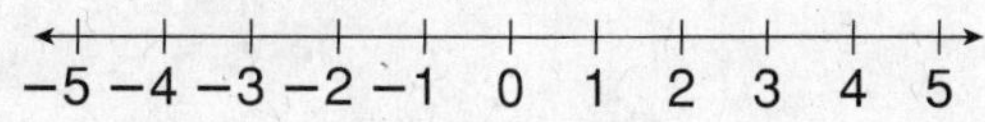

10. $x \geq 4$ or $x \leq -1$

- Draw a closed circle at −1 and a closed circle at 4.
- Shade the line to the left of −1 and to the right of 4.

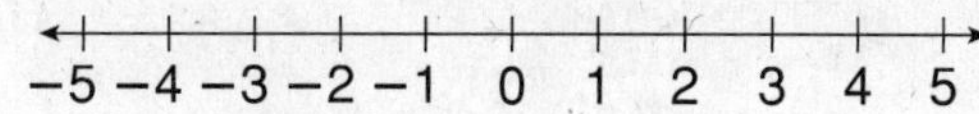

Holt Mathematics

 # Challenge
Square Equations

Some equations can be solved using square roots.

Example 1:

Solve $x^2 = 25$.

Think: $5^2 = 25$, and $(-5)^2 = 25$.

So, the equation has two solutions.

$x = -5, 5$

Example 2:

Solve $\dfrac{w}{2} = \dfrac{8}{w}$.

First, cross multiply: $w^2 = 16$.

Think: $4^2 = 16$, and $(-4)^2 = 16$.

So, the equation has two solutions.

$y = -4, 4$

Solve.

1. $x^2 = 64$

2. $y^2 = 81$

3. $n^2 = 121$

____

4. $a^2 = 49$

5. $m^2 = 100$

6. $s^2 = 169$

____

7. $\dfrac{c}{3} = \dfrac{3}{c}$

8. $\dfrac{k}{4} = \dfrac{25}{k}$

9. $\dfrac{e}{4} = \dfrac{16}{e}$

____

10. $\dfrac{9}{t} = \dfrac{t}{4}$

11. $\dfrac{16}{d} = \dfrac{d}{9}$

12. $z^2 - 1 = 35$

____

13. $p^2 - 5 = 20$

14. $h^2 + 14 = 63$

15. $29 + v^2 = 110$

____

Holt Mathematics

<table><tr><td>LESSON
12-4</td><td></td></tr></table>

Problem Solving
Inequalities

Write the correct answer.

The American College of Sports Medicine recommends exercising at an intensity of 60% to 90% of your maximum heart rate.

Heart Rates by Age

Age	Maximum Heart Rate	Target Range
20–24	200	120–180
25–29	195	117–176
30–34	190	114–171
35–39	185	111–167
40–44	180	108–162

1. Mara is 25 years old. Write a compound inequality to represent her target heart rate range while bike riding.

____

2. Leia is 38 years old. Write a compound inequality to represent the zone between her maximum heart rate and the upper end of her target range.

____

3. Rudy is 42 years old. Write an inequality to represent at least 70% of his maximum heart rate.

____

4. Write a compound inequality to represent 60% to 90% of your maximum heart rate in ten years.

____

Choose the letter for the graph that represents each statement.

5. Alena decided to pay not more than $25 to get her old bike repaired.

 A I   **C** III

 B II   **D** IV

6. It was so cold last week that the temperature never reached 25°F.

 F I   **H** III

 G II   **J** IV

7. There were at least 25 people ahead of Ivan in the cafeteria line.

 A I   **C** III

 B II   **D** IV

8. The garden yielded more than 25 pounds of potatoes.

 F I   **H** III

 G II   **J** IV

I.   0 5 10 15 20 25 30 35 40 45 50

II.   0 5 10 15 20 25 30 35 40 45 50

III.   0 5 10 15 20 25 30 35 40 45 50

IV.   0 5 10 15 20 25 30 35 40 45 50

Holt Mathematics

LESSON 12-4
Reading Strategies
Connecting Words and Symbols

An **inequality** is a comparison of two unequal values. This chart will help you understand both words and symbols for inequalities.

The team has scored fewer than 5 runs in each game. "Fewer than 5" means "less than 5." Symbol for "less than 5": < 5	No more than 8 people can ride in the elevator. "No more than 8" Means **"8 or less than 8."** Symbol for "less than or equal to 8": ≤ 8

Inequalities

More than 25 students try out for the team each year. "More than 25" means **"a number greater than 25."** Symbol for "greater than 25": > 25	There are at least 75 fans at each home game. "At least 75" means "75 or more" or **"a number greater than or equal to 75."** Symbol for "greater than or equal to 75": ≥ 75

Use the chart to answer each question.

1. What is an inequality?

____

____

2. Explain the difference between the symbols $<$ and $\leq$.

____

____

3. Explain the difference between the symbols $>$ and $\geq$.

____

____

4. Write $<$, $>$, $\leq$ or $\geq$ to describe the number of students in each homeroom: There is a limit of 30 students for each homeroom.

____

Holt Mathematics

Puzzles, Twisters & Teasers

LESSON 12-4 *Hard to Swallow!*

Find and circle words from the list in the word search (horizontally, vertically or diagonally). Find a word that answers the riddle. Circle it and write it on the line.

inequality	compound	value	symbol	graph
less	greater	equal	more	fewer

```
I N E Q U A L I T Y V F W
L Z X C V M N M K I A E N
E A S D F G O J K L L W J
S W E R K C F R T G U E K
S Y M B O L M K E U E R U
V F R G R A P H B G T N O
Y H N J E Q U A L C D E P
H J G R E A T E R A S E C
A S D F C O M P O U N D S
L I M E S T O N E N M J V
Q W E R T Y U I O P L K X
```

What kind of stone is the most sour? _______________________________

Holt Mathematics

Name ______________________ Date ____________ Class ____________

LESSON 12-5

Practice A

Solving Inequalities by Adding or Subtracting

Solve. Then match each solution set with its graph.

1. $r - 1 > 2$ ____________

2. $m + 3 \leq 6$ ____________

3. $x - 4 < -1$ ____________

4. $k + 2 \geq 5$ ____________

A.
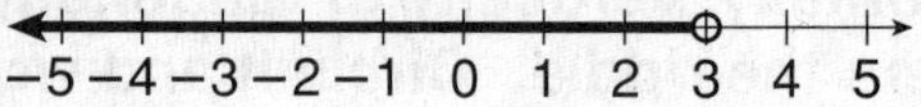

B.
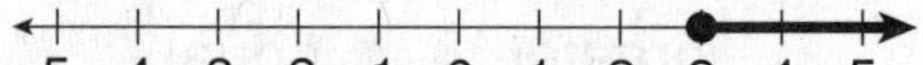

C.

D.

Solve. Check each answer.

5. $a + 7 > 2$ ____________

6. $h - 9 \leq 3$ ____________

7. $v - 8 < -16$ ____________

8. $14 + y > -7$ ____________

9. $j - 17 \geq -8$ ____________

10. $t + 25 \leq 9$ ____________

11. $w + 23 < -35$ ____________

12. $f + 39 \geq 11$ ____________

13. $54 + b \leq 45$ ____________

14. Friday's high temperature was 20°F. Saturday's high temperature was forecast to be no more than 8°F warmer than Friday's high temperature. According to the forecast, what are the possible high temperatures for Saturday?

____

LESSON 12-5

Practice B
Solving Inequalities by Adding or Subtracting

Solve. Then graph each solution set on a number line.

1. $y - 5 > -2$ _________

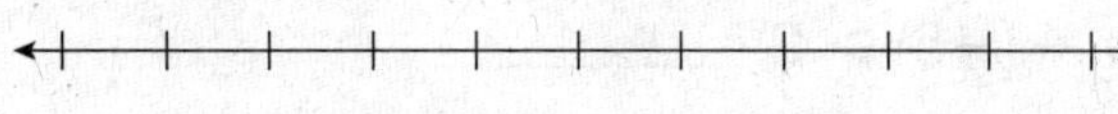

2. $n + 5 \leq 11$ _________

3. $x + 4 < -1$ _________

4. $h + 20 > 2$ _________

5. $p + 9 \geq -3$ _________

6. $s - 7 < -16$ _________

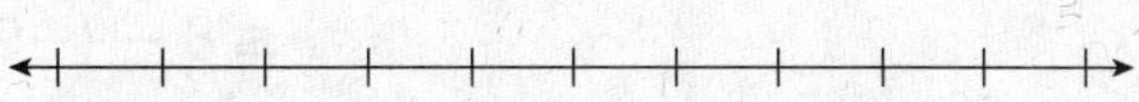

Solve. Check each answer.

7. $41 + g > 27$

8. $w + 23 \geq -18$

9. $a + 15 \leq 9$

_________ _________ _________

10. $z + 27 < 16$

11. $-3 \leq t + 17$

12. $78 \geq b + 64$

_________ _________ _________

13. In order for a field trip to be scheduled, at least 30 students must sign up. So far, 23 students have signed up. At least how many more students must sign up in order for the field trip to be scheduled?

____

Holt Mathematics

Practice C
Solving Inequalities by Adding or Subtracting

Solve. Then graph each solution set on a number line.

1. $a - (-5) > 2$ _________

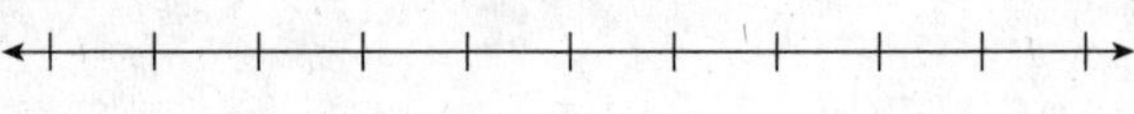

2. $m - 3 \le 7 - 11$ _________

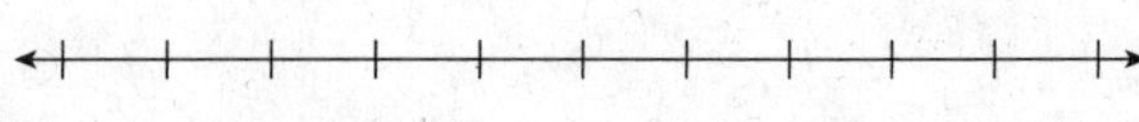

3. $x + 4 > -1 + 8$ _________

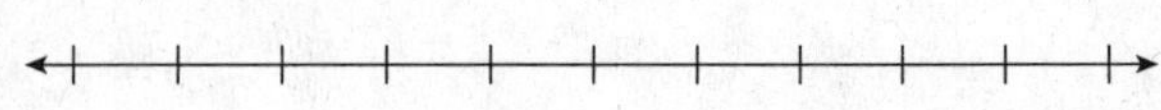

Solve. Check each answer.

4. $j - 14 + 20 > 25$

5. $w - 9 - (-1) \ge 3$

6. $r - 17 < 14 - 16$

____

7. $24 - 11 + k > 27 - 30$

8. $x - (-13) + 18 \le 8 - (-9)$

9. $9 - 15 \le b + 16$

____

10. $t - 5.3 + 2.8 < 9.5 - 11.2$

11. $s + 6.2 - 5.9 \ge 1.3 + 3.7$

____

12. $1\frac{1}{2} + z \le 2\frac{7}{8} - 4\frac{3}{4}$

13. $n + \frac{1}{6} > 1\frac{1}{3} - 2$

____

14. Nine members of the drama club are going to New York to see a Broadway play as a group. The total cost for the tickets, including a \$15.00 handling fee, will be no more than \$285.00. What is the maximum cost of each ticket? _________

Holt Mathematics

Reteach

Solving Inequalities by Adding or Subtracting

Inequalities with variables may have more than one solution. All the solutions of an inequality are called the **solution set**.

You can solve an inequality involving addition or subtraction just as you would solve an equation.

Complete the steps to solve, graph, and check the inequality.

1. $n - 13 < 7$

$n - 13 + $ _____ $< 7 + $ _____

$n < $ _____

Graph the inequality.

Check:

Pick any number in the solution set of $n < 20$.
$19 < 20$

Substitute 19 into the inequality.

$n - 13 < 7$

_____ $- 13 \overset{?}{<} 7$

_____ $\overset{?}{<} 7$ ✔

2. $x + 18 \geq 2$

$x + 18 - $ _____ $\geq 2 - $ _____

$x \geq $ _____

Graph the inequality.

Check:

Pick any number in the solution set of $x \geq -16$.
$0 \geq -16$

Substitute 0 into the inequality.

$x + 18 \geq -2$

_____ $+ 18 \overset{?}{\geq} -2$

_____ $\overset{?}{\geq} -2$ ✔

Solve. Then graph each solution set.

3. $d + 3 > -5$ __________

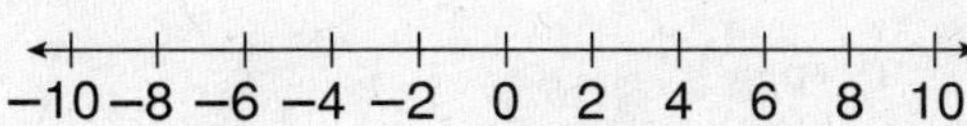

-10 -8 -6 -4 -2 0 2 4 6 8 10

4. $s - 10 < -6$ __________

-10 -8 -6 -4 -2 0 2 4 6 8 10

Solve. Check each answer.

5. $y + 9 > 20$ 　　　**6.** $h + 17 \leq -6$ 　　　**7.** $a - 4 \geq -18$

_________ 　　　_________ 　　　_________

Holt Mathematics

Challenge
Absolute Value Equations

The absolute value of a number x is:

$$|x| = x \text{ if } x > 0 \qquad \text{and} \qquad |x| = -x \text{ if } x < 0$$

When you solve an absolute value equation, you must consider both of these cases.

Example: Solve $|n + 8| = 20$.

Case 1: $n + 8 = 20$
$n = 12$

Case 2: $-(n + 8) = 20$
$-n - 8 = 20$
$-n = 28$
$n = -28$

The equation has two solutions: $n = 12, -28$

Solve.

1. $|t - 4| = 9$

2. $|6s| = 72$

3. $|y| + 3 = 7$

____

4. $|g + 6| = 3$

5. $|2m - 7| = 29$

6. $|4p - 1| = 15$

____

7. $|c + 7| = 13$

8. $|3n - 4| = 11$

9. $|2e| - 8 = 24$

____

10. $|5x| - 8 = 32$

11. $|9k| + 3 = 48$

12. $|2s - 6| = 20$

____

13. $|3w - 5| = 4$

14. $|4b - 7| = 1$

15. $|6r - 1| = 17$

____

Holt Mathematics

Problem Solving
Solving Inequalities by Adding or Subtracting

Write the correct answer.

1. A small car averages up to 29 more miles per gallon of gas than an SUV. If a small car averages 44 miles per gallon, what is the average miles per gallon for an SUV?

2. Carlos is taking a car trip that is more than 240 miles, depending on the route he chooses. He has already driven 135 miles. How much farther does he have to go?

3. Driving into the city usually takes 25 minutes. If there is a lot of traffic, the trip can take up to 45 minutes. How much additional time should you allow during a heavy traffic period?

4. To qualify for the heavyweight wrestling division, Kobe must weigh at least 180 pounds. If Kobe weighs 168 pounds now, how much weight should he gain?

Choose the letter for the best answer.

5. On one day, the range of temperatures in one state was at most 27°. If the lowest temperature in the state was 59°, what was the highest temperature?

   A $t > 86°$ **C** $t = 86°$

   B $t \le 86°$ **D** $t \ge 86°$

6. The highest possible score on the Scholastic Aptitude Test is 2,400. Rebecca scored 1780. She needs a score of at least 1,950 to qualify for a scholarship. How much higher must her score be?

   F $s \le 170$ **H** $s \ge 170$

   G $s \le 450$ **J** $s \ge 450$

7. Romero is saving to buy an Apex Model 12 Computer. The lowest price that Romero can find for the computer is $1,250. Romero now has $825. His grandmother is going to give him another $200. How much more money does Romero need?

   A $x < \$225$ **C** $x \ge \$225$

   B $x < \$425$ **D** $x \ge \$425$

8. The seating capacity of the school gym is 550. So far, there are 210 fans at a basketball game. How many more fans could attend the game?

   F $f \ge 340$

   G $f > 210$

   H $f \le 340$

   J $f < 340$

Holt Mathematics

 Reading Strategies
Use a Graphic Organizer

You can use this chart to help you understand solving two-step inequalities.

<table>
<tr><td colspan="2">

Solution of the Inequality
Any value of the variable that makes the inequality true.

</td><td colspan="2">

Solution Set
All possible solutions of an inequality.

</td></tr>
<tr><td colspan="4" align="center">

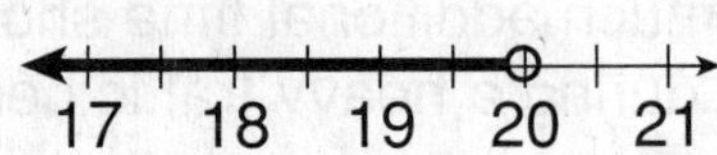

</td></tr>
<tr><td colspan="2">

Addition Inequality
$$x + 4 > 25$$
$$x + 4 - 4 > 25 - 4$$
$$x > 21$$

19 20 21 22 23 24

</td><td colspan="2">

Subtraction Inequality
$$y - 8 < 12$$
$$y - 8 + 8 < 12 + 8$$
$$y < 20$$

17 18 19 20 21

</td></tr>
</table>

Use the graphic organizer to answer the following questions.

1. What is a solution of an inequality?

____

____

2. What is the solution set?

____

____

3. What are 3 solutions for $x > 21$?

____

4. What are 3 solutions for $y < 20$?

____

5. Why is $x + 4 > 25$ called an Addition Inequality?

____

____

Holt Mathematics

<table><tr><td>**LESSON**
12-5</td><td></td></tr></table>

Puzzles, Twisters & Teasers
Are All Puzzles Created Equal?

Across

1. You can use an _____ to solve a problem.

4. Inequalities with _____ usually have many solutions.

5. Any value of a variable that makes a _____ true is a solution of the inequality.

7. When _____ your solution, you should choose a number in the solution set that is easy to work with.

9. All of the solutions of an inequality are called the _____ set.

Down

2. You can find solution sets for inequalities by _____ the variable.

3. You can check your answer by choosing any number in the solution set and _____ it into the original inequality.

6. You can solve one-_____ inequalities by adding or subtracting.

8. When you add or _____ the same number to both sides of an inequality, the statement will still be true.

Holt Mathematics

LESSON 12-6 **Practice A**
Solving Inequalities by Multiplying or Dividing

Solve. Choose the letter for the best answer.

1. $5n \leq -20$

    A $n \leq -4$ **C** $n \geq -4$

    B $n < -4$ **D** $n \geq 4$

2. $-3k > -21$

    F $k > 7$ **H** $k < 7$

    G $k > -7$ **J** $k < -7$

3. $\dfrac{m}{5} \geq -1$

    A $m \leq -5$ **C** $m \leq 5$

    B $m \geq -5$ **D** $m \geq 5$

4. $\dfrac{v}{-8} < -4$

    F $v \geq 32$ **H** $v < 32$

    G $v \leq -32$ **J** $v > 32$

5. $-2y < 12$

    A $y < 6$ **C** $y > 6$

    B $y < -6$ **D** $y > -6$

6. $\dfrac{h}{5} \leq 4.2$

    F $h \leq 21$ **H** $h \geq 21$

    G $h \leq 22$ **J** $h \geq 22$

7. $\dfrac{z}{-9} \leq -4$

8. $\dfrac{b}{3} > 15$

9. $\dfrac{m}{6} < -2.5$

___________ ___________ ___________

Solve. Check each answer.

10. $7c < -42$ **11.** $-20a \geq -100$ **12.** $-8t > 5$

___________ ___________ ___________

13. It cost \$730 to put on the school play. How many tickets must
be sold at \$6 apiece in order to make a profit?

____

Holt Mathematics

Practice B
Solving Inequalities by Multiplying or Dividing

Solve.

1. $\dfrac{n}{5} \le 1.6$

2. $\dfrac{b}{3} > -8$

3. $\dfrac{a}{3} \ge -9$

__________   __________   __________

4. $\dfrac{t}{-6} < -7$

5. $\dfrac{s}{-12} \le -5$

6. $\dfrac{r}{5.3} \le 6$

__________   __________   __________

Solve. Check each answer.

7. $8c < -64$

8. $-16a \ge -24$

9. $-12t > 9$

__________   __________   __________

10. $-3s \le -180$

11. $18b > -24$

12. $-6m \ge 4$

__________   __________   __________

13. It cost Sophia $530 to make wind chimes. How many wind chimes must she sell at $12 apiece to make a profit?

____

14. It cost the Wilson children $55 to make lemonade. How many glasses must they sell at 75¢ each to make a profit?

____

15. Jorge's soccer team is having its annual fund raiser. The team hopes to earn at least three times as much as it did last year. Last year the team earned $87. What is the team's goal for this year?

____

Holt Mathematics

Practice C

LESSON 12-6

Solving Inequalities by Multiplying or Dividing

Solve.

1. $14w \le -2.8$

2. $-0.6v > -54$

3. $\dfrac{u}{7} \ge -9.8$

____

4. $\dfrac{f}{-0.5} < -0.2$

5. $-2.5e < 1.5$

6. $\dfrac{m}{4.5} \le 6.3$

____

Solve. Check each answer.

7. $4p < (-2)^3$

8. $-8c \ge -4 \cdot 6$

9. $-12a > 8^2$

____

10. $\dfrac{x}{-4 \cdot 3} \le -5 \cdot (-3)$

11. $\dfrac{d}{6} > -3\dfrac{2}{3}$

12. $1\dfrac{1}{6}k < 14$

____

Write an inequality, then solve.

13. Four families are planning to make a joint purchase at a warehouse club. In order to purchase no more than 15 pounds of potatoes for each family, how many potatoes should they buy?

____

14. Lila's swim team is raising money to help buy a new pool cover. The team has raised $635. This is 65% of their goal. At least how much more do they need to reach their goal?

____

Holt Mathematics

 # Reteach
Solving Inequalities by Multiplying or Dividing

When you multiply or divide both sides of an inequality by the same *positive* number, the direction of the inequality symbol remains the same.

$3p < -18$

Divide both sides by *positive* 3.

$\dfrac{3p}{3} < \dfrac{-18}{3}$

$p < -6$

$\dfrac{x}{5} \geq -4$

Multiply both sides by *positive* 5.

$5 \cdot \dfrac{x}{5} \geq 5 \cdot -4$

$x \geq -20$

When you multiply or divide both sides of an inequality by the same *negative* number, the direction of the inequality symbol is reversed.

$-6y > -42$

Divide both sides by *negative* 6.

$\dfrac{-6y}{-6} < \dfrac{-42}{-6}$

$y < 7$

$\dfrac{m}{-2} \leq 15$

Multiply both sides by *negative* 2.

$-2 \cdot \dfrac{m}{-2} \geq -2 \cdot 15$

$m \geq -30$

Solve.

1. $\dfrac{a}{8} < -3$

Multiply by ___. The direction of the inequality symbol

_________________.

$8 \cdot \dfrac{a}{8} \ \square \ 8 \cdot -3$

$a \ \square \ -24$

2. $-4s \geq -36$

Divide by ___. The direction of the inequality symbol

_________________.

$\dfrac{-4s}{-4} \ \square \ \dfrac{-36}{-4}$

$s \ \square \ 9$

Solve. Check each answer.

3. $\dfrac{r}{-7} \geq 2$ **4.** $9b < -54$ **5.** $-3n > -36$ **6.** $\dfrac{c}{6} < -8$

_________   _________   _________   _________

Holt Mathematics

Challenge
LESSON 12-6 *Absolute Value Inequalities*

Use the definition of absolute value to solve absolute value inequalities.

Remember: $|x| = x$ if $x > 0$ and $|x| = -x$ if $x < 0$

Example 1: $|10 - 2x| > 6$

Case 1: $10 - 2x > 6$
$$-2x > -4$$
$$x < 2$$

Case 2: $-(10 - 2x) > 6$
$$-10 + 2x > 6$$
$$2x > 16$$
$$x > 8$$

Solution: $x < 2$ or $x > 8$

Example 2: $|2y - 3| \leq 7$

Case 1: $2y - 3 \leq 7$
$$2y \leq 10$$
$$y \leq 5$$

Case 2: $-(2y - 3) \leq 7$
$$-2y + 3 \leq 7$$
$$-2y \leq 4$$
$$y \geq -2$$

Solution: $-2 \leq y \leq 5$

Solve.

1. $|m - 6| \leq 8$

2. $|x + 4| < 10$

3. $|6 - y| \leq 2$

4. $|b + 5| > 8$

5. $|4 - k| \geq 5$

6. $|2 - p| - 3 < 1$

7. $4 - 3|e| < 1$

8. $\left|\dfrac{h}{6}\right| \geq 1$

9. $\left|\dfrac{n}{4}\right| \leq 2$

10. $|2a + 1| \geq 1$

11. $|z - 2| < 5$

12. $\left|\dfrac{c}{2} - 3\right| \leq 2$

Holt Mathematics

Problem Solving
Solving Inequalities by Multiplying or Dividing

Write the correct answer.

1. U.S. Postal Service regulations state that a package can be mailed using Parcel Post rates if it weighs no more than 70 pounds. What is the maximum number of books weighing 12 pounds each that can be mailed in one box using Parcel Post rates?

2. Marc wants to buy a set of at least 6 antique chairs for his dining room. He has decided to spend no more than $390. What is the most he can spend per chair?

3. Alfonso earns $9.00 per hour working part-time as a lab technician. He wants to earn more than $144 this week. At least how many hours does Alfonso have to work?

4. Mrs. Menendez invited 8 children to her son's birthday party. She wants to make sure that each child gets at least 4 small prizes. At least how many prizes should she buy?

Choose the letter for the best answer.

5. The Computer Club spent $2,565 on mouse pads. The members plan to sell the mouse pads during the book fair. If they charge $9.50 for each mouse pad, how many must they sell in order to make a profit?

   A $n \leq 271$ C $n \geq 271$

   B $n \leq 270$ D $n \geq 270$

6. The Parents Organization bought 1,000 bumper stickers at $1.25 each to sell at football games. They want to make at least $750 profit. What should be the selling price of the bumper stickers?

   F $p \geq \$1.25$ H $p \geq \$1.75$

   G $p \geq \$1.50$ J $p \geq \$2.00$

7. In 2000, the national ratio of students to computers with Internet access in public schools was 7:1. Winston School had 623 students. If the school had a lower ratio, how many computers with Internet access did Winston School have?

   A $c \geq 90$ C $c \geq 91$

   B $c \leq 90$ D $c \leq 91$

8. A new theme park averaged fewer than 2,000 visitors per week during the winter months. What was the average daily attendance?

   F $a \geq 290$ H $a > 385$

   G $a \leq 186$ J $a < 286$

Holt Mathematics

LESSON 12-6
Reading Strategies
Analyze Information

Look at the steps for solving multiplication and division inequalities.

Steps for Solving an Inequality	Division Inequality	Multiplication Inequality
Step 1: Isolate the variable.	$\frac{x}{8} < 5$	$4y > 12$
Step 2: Multiply both sides of a division equation by the same number. Divide both sides of a multiplication equation by the same number.	$\frac{x}{8} \cdot 8 < 5 \cdot 8$	$\frac{4y}{4} > \frac{12}{4}$
Step 3: Compute to solve.	$x < 40$	$y > 3$

Use the chart to answer the following questions.

1. What do you do to get the variable by itself in a division inequality?

____

2. What was $\frac{x}{8}$ multiplied by in the example?

____

3. What is the solution of the division inequality?

____

4. What are three numbers that make the division inequality true?

____

5. What do you do to get the variable by itself in a multiplication inequality?

____

6. What was $4y$ divided by in the multiplication inequality?

____

7. What is the solution to the multiplication inequality?

____

8. What are three numbers that are part of the solution?

____

Holt Mathematics

Puzzles, Twisters & Teasers

LESSON 12-6 *Bone Tired!*

You dog has lost his bone and wants to find it before he takes a nap. Circle the correct symbol to solve each inequality. Then use the number under the correct answer to move through the maze. The first one is done for you.

1. $5w$ __ 517 $w = 111$ ⊙ $<$ Move ____ spaces right

    $555 > 517$ 2   5

2. $3m$ __ -15 $m = 2$ $>$ $<$ Move ____ spaces down

                               8   3

3. $-12y$ __ -60 $y = 3$ $>$ $<$ Move ____ spaces right

                               9   3

4. $8y$ __ 16 $y = 1$ $>$ $<$ Move ____ spaces up

                               8   6

5. $4x$ __ 9 $x = 5$ $>$ $<$ Move ____ spaces right

                               3   7

6. $25c$ __ 200 $c = 4$ $>$ $<$ Move ____ spaces up

                               4   2

7. $-6r$ __ -48 $r = 12$ $>$ $<$ Move ____ spaces right

                               9   5

8. $\dfrac{x}{11}$ __ 3 $x = 22$ $>$ $<$ Move ____ spaces down

                               3   8

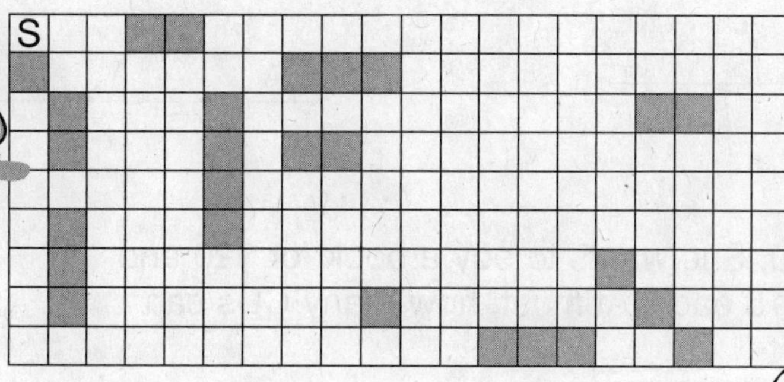

Holt Mathematics

LESSON 12-7 — Practice A
Solving Two-Step Inequalities

Solve. Cross out each inequality in the box that matches a solution. Then graph each solution set.

$$x > 8 \qquad x < -8 \qquad x \geq -8 \qquad x < 8 \qquad x \geq 8 \qquad x \leq -8$$

1. $3x - 5 < 19$ ___________

2. $-2x + 12 < -4$ ___________

3. $\dfrac{x}{4} + 7 \geq 9$ ___________

4. $\dfrac{x}{-2} - 3 \geq 1$ ___________

Solve. Then graph each solution set.

5. $7y - 8 > 6$ ___________

6. $-4d + 15 \leq -1$ ___________

7. $\dfrac{r}{-6} + 5 < 7$ ___________

8. Margie has $100. She wants to buy a book for $20 and some CDs for $15 each. At most, how many CDs can Margie buy?

Holt Mathematics

Practice B
Solving Two-Step Inequalities

Solve. Then graph each solution set on a number line.

1. $5x - 8 < 17$ __________

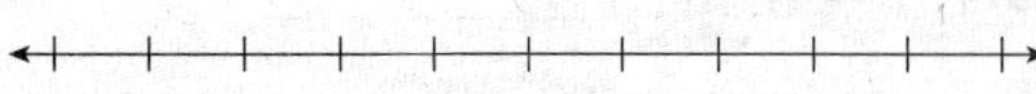

2. $\dfrac{r}{3} + 5 \geq 9$ __________

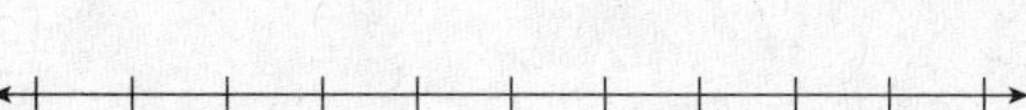

3. $-4n + 8 < -4$ __________

4. $\dfrac{z}{7} - 6 \geq -5$ __________

5. $\dfrac{w}{-5} + 4 < 9$ __________

6. $\dfrac{u}{2} - 5 \leq -9$ __________

Solve.

7. $-7d + 8 > 29$

8. $4g - 18 \leq -2$

9. $12 - 3b < 9$

____

____

____

10. $\dfrac{a}{-4} - 7 < -2$

11. $9 + \dfrac{c}{6} \leq 17$

12. $-\dfrac{2}{3}p - 8 \geq 4$

____

____

____

13. Fifty students in the seventh grade are trying to raise at least \$2,000 for sports supplies. They have already raised \$750. How much should each student raise, on average, in order to meet the goal?

Holt Mathematics

<table>
<tr><td>LESSON
12-7</td><td>

Practice C
Solving Two-Step Inequalities
</td></tr>
</table>

Solve. Then graph each solution set on a number line.

1. $5y + 1 - 8 < 13$ __________

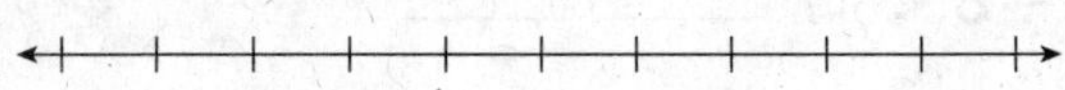

2. $\dfrac{r}{8} - 5 - 6 \geq -9$ __________

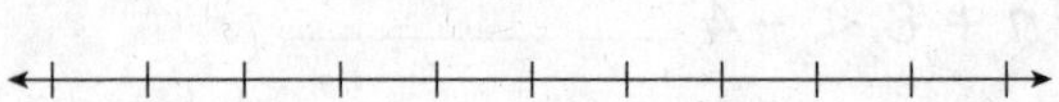

3. $-7m - 6 < -4 - 9$ __________

Solve.

4. $\dfrac{s}{5} - 6 + 3 \geq -5$ **5.** $\dfrac{w}{-2} + 15 < 9 - (-3)$ **6.** $\dfrac{u}{3} - 0.5 \leq -9.5$

__________   __________   __________

7. $-4d + 6 - d \leq 21$ **8.** $6p - 18 - 2p \leq -2$ **9.** $9h - 20 - 3h < -8$

__________   __________   __________

10. $\dfrac{a}{2} - 7\dfrac{1}{2} + \dfrac{a}{2} < -2\dfrac{1}{2}$ **11.** $4.5 + \dfrac{c}{2.3} \leq 6.1$ **12.** $-\dfrac{2}{3}q - \dfrac{3}{4} \geq \dfrac{1}{12}$

__________   __________   __________

13. Max earns $1,800 per month, plus a commission of 3%
of his sales. He wants to earn at least $3,000 this month.
What amount of sales does Max need this month?

____

Holt Mathematics

Reteach
Solving Two-Step Inequalities

Two-step inequalities can be solved by first undoing addition or subtraction, then undoing multiplication or division.

Complete the steps to solve the inequality. Then graph the solution set.

1. $-3x + 35 > -10$

$-3x + 35$ _____ > -10 _____ ← First undo addition or subtraction.

$-3x >$ _____ ← Then undo multiplication or division.

$\dfrac{-3x}{-3} \ \square \ \dfrac{}{-3}$ ← Divide by -3. The inequality symbol is reversed.

$x \ \square$ _____

Graph the inequality.

$-10 \ -5 \ \ 0 \ \ 5 \ \ 10 \ \ 15 \ \ 20 \ \ 25 \ \ 30 \ \ 35 \ \ 40$

Solve. Then graph each solution set.

2. $7x + 32 < 18$ _____

$-5 \ -4 \ -3 \ -2 \ -1 \ \ 0 \ \ 1 \ \ 2 \ \ 3 \ \ 4 \ \ 5$

3. $\dfrac{t}{4} - 8 \le -5$ _____

$0 \ \ 2 \ \ 4 \ \ 6 \ \ 8 \ \ 10 \ \ 12 \ \ 14 \ \ 16 \ \ 18 \ \ 20$

4. $6f - 6 > 48$ _____

$2 \ \ 3 \ \ 4 \ \ 5 \ \ 6 \ \ 7 \ \ 8 \ \ 9 \ \ 10 \ \ 11 \ \ 12$

5. $-4w - 13 \ge 15$ _____

$-9 \ -8 \ -7 \ -6 \ -5 \ -4 \ -3 \ -2 \ -1 \ \ 0 \ \ 1$

6. $\dfrac{k}{-5} - 6 < -9$ _____

$-10 \ -5 \ \ 0 \ \ 5 \ \ 10 \ \ 15 \ \ 20 \ \ 25 \ \ 30 \ \ 35 \ \ 40$

7. $7 - 2p > -5$ _____

$-10 \ -8 \ -6 \ -4 \ -2 \ \ 0 \ \ 2 \ \ 4 \ \ 6 \ \ 8 \ \ 10$

Holt Mathematics

Challenge
Inequalities With Two Variables

You can determine if an ordered pair is a solution of an inequality by substituting the values into the inequality.

- If the inequality is true, the ordered pair is a solution of the inequality.
- If the inequality is false, the ordered pair is not a solution of the inequality.

Example: Is $(1, 1)$ or $(2, -1)$ a solution of the inequality $y < 3x - 2$?

$(1, 1)$ $\qquad$ $y < 3x - 2$ $\qquad\qquad$ $(2, -1)$ $\qquad$ $y < 3x - 2$

$\qquad\qquad\quad$ $1 < 3(1) - 2$ $\qquad\qquad\qquad\qquad$ $-1 < 3(2) - 2$

$\qquad\qquad\quad$ $1 < 1$ $\;$ False $\qquad\qquad\qquad\qquad\;$ $-1 < 4$ $\;$ True

$\qquad\qquad\quad$ $(1, 1)$ is not a solution. $\qquad\qquad$ $(2, -1)$ is a solution.

Solve.

1. Which of the following ordered pairs are solutions of $y < 4x - 3$:
   $(-1, 1)$, $(2, 5)$, $(-\frac{1}{2}, -7)$, $(0, -4)$?

   ____

2. Which of the following ordered pairs are solutions of $y \geq 6 - 2x$:
   $(0, 0)$, $(1, 5)$, $(3, 6)$, $(\frac{1}{2}, 8)$?

   ____

3. Which of the following ordered pairs are solutions of $4x + 2y \leq 20$:
   $(3, 0)$, $(0, 12)$, $(5, 5)$, $(2, -4)$?

   ____

4. For the inequality $y < 5x - 2$, let $x = 0$. Solve for y.

   ____

5. List 4 ordered pairs that are solutions to $y < 5x - 2$.

   ____

6. List 4 ordered pairs that are solutions of $3x + 2y > 4$.

   ____

Holt Mathematics

 # Problem Solving
Solving Two-Step Inequalities

Write the correct answer.

1. Grace earns $7 for each car she washes. She always saves $25 of her weekly earnings. This week, she wants to have at least $65 in spending money. At least how many cars must she wash?

2. Monty has saved $400 to spend on a video game player and games. The player he wants costs $275. The games each cost $39. At most, how many games can he buy along with the player?

3. A video game club charges $8 per month as a membership fee, plus $2.75 for each game rental. Eugenie plans to join and rent no more than 5 games a month. What amount should she budget each month for video games?

4. Cooper Middle School has a goal of collecting more than 1,000 cans of food in a food drive. So far, 375 cans have been collected. During the last 13 days of the drive, at least how many cans must be collected each day in order to meet the goal?

Choose the letter for the best answer.

5. In January 2005, it cost $0.37 to mail a letter weighing up to 1 ounce. Each additional ounce or part of an ounce cost $0.23. At most, what is the weight of a letter with $1.06 in postage?

A $w < 4$ oz   **C** $w \le 4$ oz

B $w < 3$ oz   **D** $w \le 3$ oz

6. Martin is planning a hedge along the back of his yard. The total length can be no more than 23 feet, and he will put a 4-foot-wide gate in the hedge. Each plant needs 2.5 feet of space to grow properly. How many plants should he buy?

F $n \le 7$ **H** $n \le 9$

G $n < 7$ **J** $n < 9$

7. The 12 members of the Middle School filmmaking club need to raise at least $1,400 to make a short film. They already have raised $650. How much more should each member raise on average?

A $x \ge \$62.50$ **C** $x < \$62.50$

B $x \le \$62.50$ **D** $x > \$62.50$

8. The rule of thumb in filmmaking is that you must shoot at least 3 minutes of film for every minute in a movie's "final cut." A 30-minute roll of film costs $250. How much will film cost to make a 90-minute movie?

F $c \ge \$22,500$ **H** $c \le \$67,500$

G $c \le \$7,500$ **J** $c \ge \$2,250$

Holt Mathematics

Reading Strategies

LESSON 12-7 *Use a Flowchart*

You can use a flowchart to solve a two-step inequality.

$4x + 5 < 13$

> **Step 1:** Combine the terms without variables. Use the inverse operation.
>
> $4x + 5 - 5 < 13 - 5$
>
> $4x < 8$

> **Step 2:** Get the variable by itself. Use the inverse operation.
>
> $\dfrac{4x}{4} > \dfrac{8}{4}$

> **Step 3:** Simplify.
>
> $x < 2$

> **Step 4:** Graph.
>
> 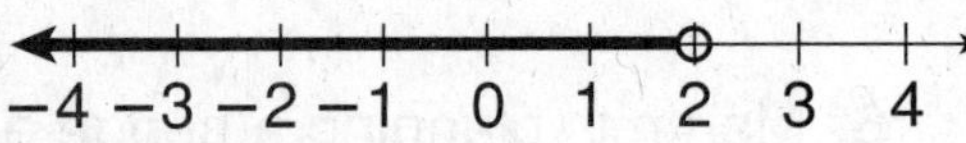

Use the chart to answer the following questions.

1. What is the first step in the flowchart?

____

____

2. Why is 5 subtracted from both sides of the inequality?

____

____

3. What is the second step in the flowchart?

____

____

4. What operation is used to get *x* by itself?

____

Holt Mathematics

Puzzles, Twisters & Teasers

1-800-HELP!

Solve each inequality. Graph each solution set on the number line. Then use the letter next to your answer to solve the riddle.

1. $\frac{x}{5} - 6 < 19$

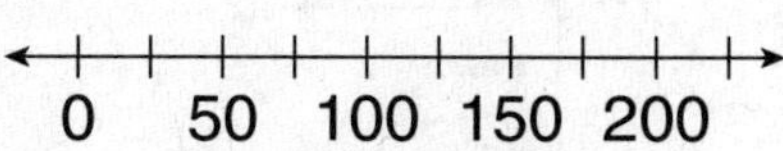

With what number did you start the graph? _________ = **H**

2. $\frac{y}{6} + 5 \leq -13$

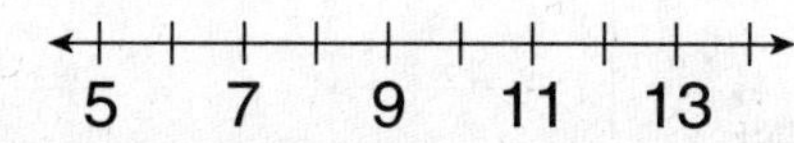

Is your circle open or closed? _________ = **E**

3. $-8x + 5 \leq -51$

With what number did you start? _________ = **I**

4. $5y - 4 > -9$

Is your circle open or filled? _________ = **B**

5. $\frac{x}{-3} + 8 > 11$

Is it possible for the solution to include -9? _________ = **N**

6. $7y - 6 \geq 22$

Is it possible for the solution to include 4? _________ = **L**

Why were the alien's eyes so big?

He saw his P ___ O ___ ___
                125   no   ●

___ ___ L ___
O   7   yes

Solve each equation. Cross out each number in the box that matches a solution.

| 6 | −8 | −4 | −6 | −18 | −3 | 2 | 4 | 8 | 3 | 18 |

1. $5x + 8 = 23$

$x = 3$

2. $-2p - 4 = 2$

$p = -3$

3. $6a - 11 = 13$

$a = 4$

4. $4n + 12 = 4$

$n = -4$

5. $9g + 2 = 20$

$g = 2$

6. $\frac{k}{6} + 8 = 5$

$k = -18$

7. $\frac{s}{3} - 4 = 2$

$s = 18$

8. $\frac{c}{2} + 5 = 1$

$c = -8$

9. $9 + \frac{a}{6} = 8$

$a = -6$

Solve. Check each answer.

10. $3v - 12 = 15$

$v = 9$

11. $8 + 5x = -2$

$x = -2$

12. $\frac{d}{4} - 9 = -3$

$d = 24$

13. An electrician charges $50 to come to your house. He also charges $25 for each hour he spends at your house. The electrician charges you a total of $125. How many hours did he spend at your house?

3 hours

3   **Holt Mathematics**

Solve. Check each answer.

1. $7x + 8 = 36$

$x = 4$

2. $-3y - 7 = 2$

$y = -3$

3. $4a - 13 = 19$

$a = 8$

4. $6a - 4 = -2$

$a = \frac{1}{3}$

5. $5k + 2 = 6$

$k = \frac{4}{5}$

6. $9m - 14 = -8$

$m = \frac{2}{3}$

Solve.

7. $\frac{v}{4} - 3 = 5$

$v = 32$

8. $\frac{u}{5} + 3 = 1$

$u = -10$

9. $6 + \frac{z}{9} = 9$

$z = 27$

10. $-7 + \frac{f}{2} = -1$

$f = 12$

11. $9 + \frac{w}{4} = -5$

$w = -56$

12. $\frac{e}{7} - 3 = -5$

$e = -14$

13. $-8 + \frac{d}{5} = 2$

$d = 50$

14. $\frac{u}{5} + 3 = 6$

$u = 15$

15. $\frac{f}{-3} + 5 = 8$

$f = -9$

16. Two years of local Internet service costs $685, including the installation fee of $85. What is the monthly fee?

$25 per month

4   **Holt Mathematics**

Solve. Check each answer.

1. $15h - 4 = 6$

$h = \frac{2}{3}$

2. $25k + 12 = -13$

$k = -1$

3. $8m - 14 = -7$

$m = \frac{7}{8}$

4. $-5m - 6 = 84$

$m = -18$

5. $3f + 8 = 28$

$f = 6\frac{2}{3}$

6. $5z + 6 = -20$

$Z = -5\frac{1}{5}$

Solve.

7. $\frac{d}{14} - 9 = 5$

$d = 196$

8. $\frac{a}{35} + 20 = 18$

$a = -70$

9. $\frac{3}{4} - 9p = -\frac{3}{5}$

$p = \frac{3}{20}$

10. $-2\frac{2}{3} + \frac{d}{6} = \frac{2}{3}$

$d = 20$

11. $-\frac{1}{4} + \frac{f}{8} = \frac{1}{2}$

$f = 6$

12. $-\frac{2}{3} = 6 + \frac{v}{-3}$

$v = 20$

Translate each equation into words, and then solve the equation.

13. $7 + \frac{n}{2} = 10$

7 more than one half a number equals 10; $n = 6$

14. $4w - 9 = -5$

4 times a number, minus 9, equals −5; $w = 1$

15. A taxi charges $1.50 plus a fee of $0.60 for each mile traveled. If a ride costs $5.40, how many miles was the ride?

$6\frac{1}{2}$ miles

5   **Holt Mathematics**

You can solve two-step equations by undoing one operation at a time. First undo any addition or subtraction, then undo any multiplication or division.

Complete the steps to solve each equation.

1.

$$7x + 3 = 31$$

$7x + 3 - \underline{3} = 31 - \underline{3}$ ← Subtract $\underline{3}$ from both sides to undo addition.

$$7x = 28$$

$\frac{7x}{7} = \frac{28}{7}$ ← Divide both sides by $\underline{7}$ to undo multiplication.

$$x = 4$$

Check

$$7x + 3 = 31$$

$7(\underline{4}) + 3 \overset{?}{=} 31$ ← Substitute $\underline{4}$ for x.

$\underline{28} + 3 \overset{?}{=} 31$

$31 \overset{?}{=} 31$ ✔ ← 4 is a solution.

2.

$$\frac{n}{6} - 8 = 4$$

$\frac{n}{6} - 8 + \underline{8} = 4 + \underline{8}$

$$\frac{n}{6} = 12$$

$6 \cdot \frac{n}{6} = \underline{6} \cdot 12$

$$n = \underline{72}$$

3.

$$8a - 5 = 11$$

$8a - 5 + \underline{5} = 11 + \underline{5}$

$$8a = \underline{16}$$

$\frac{8a}{8} = \frac{16}{8}$

$$a = \underline{2}$$

4.

$$9 + \frac{w}{2} = 12$$

$9 - \underline{9} + \frac{w}{2} = 12 - \underline{9}$

$$\frac{w}{2} = \underline{3}$$

$2 \cdot \frac{w}{2} = \underline{2} \cdot 3$

$$w = \underline{6}$$

Solve.

5. $4n + 11 = 27$

$n = 4$

6. $\frac{z}{7} - 6 = 3$

$z = 63$

7. $3 - 2k = -7$

$k = 5$

6   **Holt Mathematics**

Challenge
Consecutive Number Search

Consecutive numbers are numbers that come one after the other.

For example, 3 and 4 are consecutive whole numbers.
    6 and 8 are consecutive even numbers.
    7 and 9 are consecutive odd numbers.

If n = a number, then $n + 1$ is the next consecutive whole number after n.

If n = an even number, then $n + 2$ is the next consecutive even number after n.

If n = an odd number, then $n + 2$ is the next consecutive odd number after n.

Solve. (Hint: Let n = the first number. Then write an equation for each sum.)

1. The sum of two consecutive numbers is 73. What are the numbers?

   $n + (n + 1) = 73$; 36 and 37

2. The sum of two consecutive numbers is 131. What are the numbers?

   $n + (n + 1) = 131$; 65 and 66

3. The sum of two consecutive even numbers is 70. What are the numbers?

   $n + (n + 2) = 70$; 34 and 36

4. The sum of two consecutive even numbers is 158. What are the numbers?

   $n + (n + 2) = 158$; 78 and 80

5. The sum of two consecutive odd numbers is 108. What are the numbers?

   $n + (n + 2) = 108$; 53 and 55

6. The sum of two consecutive odd numbers is 180. What are the numbers?

   $n + (n + 2) = 180$; 89 and 91

7

Problem Solving
Solving Two-Step Equations

Write the correct answer.

1. Last week, Carlie had several rice cakes and 3 granola bars as snacks. The snacks contained a total of 800 calories. If each granola bar had 120 calories and each rice cake had 40 calories, how many rice cakes did she have?

   11 rice cakes

2. Jo eats 2,200 calories per day. She eats 450 calories at breakfast and twice as many at lunch. If she eats three meals with no snacks, which meal will contain the most calories?

   lunch

3. Erika is following a 2,200 calorie-per-day diet. She eats the recommended 9 servings of breads and cereals, averaging 120 calories per serving. She also eats 5 servings of vegetables. If the rest of her daily intake is 870 calories, what is the average number of calories in each serving of vegetables?

   50 calories

4. Brandon follows a 2,800 calorie-per-day diet. He has 11 servings of breads and cereals, which average 140 calories each. Yesterday, he had a combined 9 servings of fruits and vegetables, averaging 60 calories each. How many 180-calorie servings of meat and milk did he have to complete his diet?

   4 servings

Choose the letter for the best answer.

The table shows calories burned by a person performing different activities.

5. Kamisha swims for 0.25 hour. How many calories does she burn?

   A 30 calories   C 1.95 calories
   B 195 calories   **D** 117 calories

6. Stu jogs at a rate of 5 mi/h. How far must he jog to burn 418.5 calories?

   F 9 mi   **H** 3.75 mi
   G 4.65 mi   J 45 mi

7. Terry rides her bike for 40 minutes and plays basketball for an hour. How many calories does she burn?

   A 67 calories   **C** 670 calories
   B 560 calories   D 1,300 calories

Calories Used in Activities

Activity	Calories (per min)
Basketball	7.5
Cycling (10 mi/h)	5.5
Jogging	9.3
Swimming	7.8

8. How many hours would you have to ride your bike at 10 mi/h to burn 550 calories?

   F 1.67 hr   H 1.0 hr
   G 1.5 hr   J 0.75 hr

8

Reading Strategies
Follow a Procedure

To solve two-step equations, follow these steps.

To Solve Two-Step Equations

$3n + 5 = 23$

Step 1: Get the variable term by itself. Use the inverse operation.

$3n + 5 - 5 = 23 - 5$ Subtract 5 from both
$3n = 18$ sides.

Step 2: Get the variable by itself. Use the inverse operation.

$\dfrac{3n}{3} = \dfrac{18}{3}$ Divide both sides by 3.

Step 3: Compute and simplify the solution.

$n = 6$

Answer the following questions.

1. What is the first step in solving a two-step equation?

   Use the opposite operation to get the term with a variable by itself.

2. Which term in the equation above does not contain a variable?

   +5

3. What operation was performed to remove that term?

   5 was subtracted from both sides of the equation

4. What is the second step in solving a two-step equation?

   Get the variable by itself.

5. Which term in the equation contains a variable?

   $3n$

6. What operation was performed to get the n by itself?

   division

7. What is the third step in a two-step equation?

   Compute and simplify.

9

Puzzles, Twisters & Teasers
Sun, Sand, and Snakes!

Solve each equation. Then use the letters of the variables to answer the riddle.

1. $\dfrac{s}{-6} - 21 = -28$ $s = $ 42

2. $-4n - 7 = 17$ $n = $ −6

3. $6a + 4 = 10$ $a = $ 1

4. $\dfrac{i}{3} - 10 = -6$ $i = $ 12

5. $7t - 8 = 27$ $t = $ 5

6. $\dfrac{c}{8} + 25 = 20$ $c = $ −40

7. $-9e - 15 = 93$ $e = $ −12

8. $-3o - 8 = -35$ $o = $ 9

9. $5r + 13 = 15$ $r = $ $\dfrac{2}{5}$

10. $6m + 4 = 22$ $m = $ 3

What kind of shoes do snakes wear to the beach?

W **A T E R**   **M O C C A S I N S**
  1   5   −12 $\frac{2}{5}$ 3   9   −40   −40   1   42   12   −6   42

10

61

Practice A
Solving Multi-Step Equations

Solve. Choose the letter for the best answer.

1. $12n - 7 - 3n = 11$

2. $3a + 2 + 4a = 23$

- **A** $n = 1$
- **(B)** $n = 2$
- **C** $n = 9$
- **D** $n = 18$
- **(F)** $a = 3$
- **G** $a = 5$
- **H** $a = 7$
- **J** $a = 9$

3. $4(y + 5) + 3 = 35$

4. $2(h - 6) + 20 = -4$

- **(A)** $y = 3$
- **B** $y = 4.5$
- **C** $y = 20$
- **D** $y = 27$
- **F** $h = 4$
- **G** $h = 1$
- **(H)** $h = -6$
- **J** $h = -18$

Solve.

5. $x + 6 + 2x = 15$

6. $10b + 9 - 3b = 2$

7. $5n - 2 + 3n = 5$

$x = 3$ $b = -1$ $n = \dfrac{7}{8}$

8. $6(w - 8) + 16 = 4$

9. $2(z + 5) + 4 = -12$

10. $1.5(b + 6) + 9 = 24$

$w = 6$ $z = -13$ $b = 4$

11. Jose ran twice as many kilometers as Karen. Adding 8 to the number of kilometers Jose ran and dividing by 4 gives the number of kilometers Maria ran. Maria ran 3 kilometers. How many kilometers did Karen run?

2 kilometers

11   **Holt Mathematics**

Practice B
Solving Multi-Step Equations

Solve.

1. $15x - 8 - 3x = 16$

2. $5n + 3 + 4n = 30$

3. $h - 6 + 7h = 42$

$x = 2$ $n = 3$ $h = 6$

4. $-3g + 6 + 2g = 15$

5. $-2b + 7 - 3b = 2$

6. $5y + 1 + 3y = -15$

$g = -9$ $b = 1$ $y = -2$

7. $4k - 14 + 3k = 21$

8. $9m + 10 - 14m = -5$

9. $-2d + 18 - 4d = 60$

$k = 5$ $m = 3$ $d = -7$

10. $3(n + 5) + 2 = 26$

11. $4 - 2(v - 6) = -8$

12. $1.4 - 1.6(t + 6) = 4.6$

$n = 3$ $v = 12$ $t = -8$

13. $2.4(m - 3) + 3.8 = -8.2$ 14. $6 = 8\left(s - \dfrac{3}{4}\right) - 20$ 15. $5\left(c + \dfrac{4}{5}\right) + 6 = 50$

$m = -2$ $s = 4$ $c = 8$

16. Joel has twice as many CDs as Mariella has. Subtracting 7 from the number of CDs Joel has and dividing by 3 equals the number of CDs Blake has. If Blake has 25 CDs, how many CDs does Mariella have?

Mariella has 41 CDs.

12   **Holt Mathematics**

Practice C
Solving Multi-Step Equations

Solve.

1. $-31x - 18 + 3x = 38$

2. $15n + 16 + 4n = 130$

3. $h - 48 + 17h = 42$

$x = -2$ $n = 6$ $h = 5$

4. $-6g + 7g - 7 = 9$

5. $16 - 12a - 13a = -9$

6. $8y + 43 - 7y = -15$

$g = 16$ $a = 1$ $y = -58$

7. $4(c - 3) - 6 = 10$

8. $3.8(s - 6) + 4.6 = -22$

9. $18\left(m + \dfrac{4}{9}\right) - 16 = 118$

$c = 7$ $s = -1$ $m = 7$

10. $\dfrac{17u - 24}{18} = \dfrac{5}{9}$

11. $\dfrac{0.4n - 1}{7} = 5$

12. $\dfrac{18 - 4t}{0.5} = 20$

$u = 2$ $n = 90$ $t = 2$

13. $\dfrac{1.4m - 4.5}{1.5} = 8.2$

14. $\dfrac{\frac{1}{3}s - 2}{5} = -2$

15. $\dfrac{5.6 - 2a + 1.4}{1.2} = 4.5$

$m = 12$ $s = -24$ $a = 0.8$

16. There are 5 times as many tomato plants in the garden as there are cucumber plants. The number of pepper plants is 7 more than the sum of the number of tomato and cucumber plants. There are 19 pepper plants in the garden. How many cucumber plants are there? How many tomato plants are there?

There are 2 cucumber plants and 10 tomato plants.

13   **Holt Mathematics**

Reteach
Solving Multi-Step Equations

Some equations require more than two steps to solve. To make these equations easier to solve, first combine any like terms. Also, you may need to use the Distributive Property to complete multiplication that involves parentheses.

Complete the steps to solve each equation.

1. $5a - 7 + 3a = 17$

   $\underline{8a} - 7 = 17$ ←— Combine like terms: $5a + 3a = \underline{8a}$.

   $8a - 7 + \underline{7} = 17 + \underline{7}$ ←— Add $\underline{7}$ to both sides of the equation.

   $8a = \underline{24}$

   $\dfrac{8a}{8} = \dfrac{24}{8}$ ←— Divide both sides of the equation by $\underline{8}$.

   $a = \underline{3}$

2. $4(x + 6) - 15 = 21$

   $\underline{4}(x) + \underline{4}(6) - 15 = 21$ ←— Multiply each term in the parentheses by $\underline{4}$.

   $4x + 24 - 15 = 21$

   $4x + \underline{9} = 21$ ←— Combine like terms: $24 - 15 = \underline{9}$.

   $4x + 9 - \underline{9} = 21 - \underline{9}$ Subtract from both sides of the equation by $\underline{9}$.

   $4x = \underline{12}$

   $\dfrac{4x}{4} = \dfrac{12}{4}$ ←— Divide both sides of the equation by $\underline{4}$.

   $x = \underline{3}$

Solve.

3. $9b + 6 - 5b = 18$

4. $2y - 7 + y = 8$

5. $-5c - 12 - 5c = 8$

$b = 3$ $y = 5$ $c = -2$

6. $2(r - 3) + 6 = 14$

7. $8 + 5(s - 1) = 18$

8. $6(x + 3) + 38 = 2$

$r = 7$ $s = 3$ $x = -9$

14   **Holt Mathematics**

Challenge
Solve the System

When equations have two variables, you can sometimes find the point where two equations have the same solution—that is, where one value of x and one value of y solve both equations. This is called "solving a system of equations."
To solve a system of equations, you can substitute from one equation into the other.

Example: Solve $x + y = 6$ and $x - y = 2$.

Step 1: Solve one equation for y. $x + y = 6$
$$y = 6 - x$$

Step 2: Substitute that value of y into the second equation.
Solve for x. $x - y = 2$
$$x - (6 - x) = 2$$
$$x - 6 + x = 2$$
$$2x - 6 = 2$$
$$2x = 8$$
$$x = 4$$

Step 3: Substitute the solution for x from Step 2 into the other equation. $x + y = 6$
$$4 + y = 6$$
$$y = 2$$

Step 4: Write the solution as an ordered pair (x, y).
Solution: $(4, 2)$

Solve each system of equations.

1. $y = 3x$
   $x + y = 20$

   __(5, 15)__

2. $m = 4n$
   $m + n = 10$

   __(8, 2)__

3. $d = c - 2$
   $c + d = 12$

   __(7, 5)__

4. $v - w = 2$
   $3w + v = 14$

   __(5, 3)__

5. $j = k + 3$
   $j + k = 7$

   __(5, 2)__

6. $b = a - 9$
   $a + b = 3$

   __(6, -3)__

7. $r = 2 - s$
   $2s + r = 9$

   __(-5, 7)__

8. $e + f = 2$
   $3e + f = 8$

   __(3, -1)__

15

Problem Solving
Solving Multi-Step Equations

Write the correct answer.
To convert a temperature from degrees Fahrenheit to degrees Celsius, you can use the formula $(°F - 32)0.56 = °C$.

1. The record high temperature in North Carolina is 110°F. What is the record high in degrees Celsius?

   __43.68°C__

2. The record low temperature in Florida is −2°F. What is the record low in degrees Celsius?

   __−19.04°C__

3. The record high temperature in the United States is 134°F. This was recorded in Greenland Ranch, California, on July 10, 1913. What is that temperature in degrees Celsius?

   __57.12°C__

4. The record high in Texas is 120°F. The record low in Texas is −23°F. In degrees Celsius, what is the range between the record high and low temperatures in Texas?

   __62.16°C__

5. When the temperature is 4°C, you need to wear a heavy coat. Write 4°C as degrees Fahrenheit.

   __39.14°F__

6. When the temperature is 28°C, you might want to go to the beach. Write 28°C as degrees Fahrenheit.

   __82°F__

Choose the letter for the best answer.

7. Faith spent $78 at Fashion Warehouse. She bought 2 shirts that each cost $17.50 and a pair of shoes. How much did she spend for the shoes?
   A $34.00
   (B) $43.00
   C $60.50
   D $113.00

8. Three friends each pay $4.15 to buy a pizza. A basic pizza costs $9.45. Additional toppings cost $1 each. How many toppings were on the pizza?
   F 2 toppings
   (G) 3 toppings
   H 4 toppings
   J 5 toppings

9. Todd buys 3 CDs at $16.99 each and a DVD that costs $24.99. He pays with a $100 bill. How much change does he receive?
   (A) $24.04
   B $8.04
   C $49.03
   D $67.96

10. Marina bought 4 books. José bought half as many books as Ben bought. Together, the 3 friends bought 13 books. How many books did Ben buy?
    F 9 books
    (G) 6 books
    H 3 books
    J 2 books

16

Reading Strategies
Analyze Information

Look at the terms in this equation.

Term Term Term Term
$$4x \quad + 6 \quad + 2x \quad - 3 \quad = 33$$

Answer each question.

1. How many terms does this equation have?

   __4 terms__

2. How many terms have variables?

   __2 terms__

3. How many terms do not have variables?

   __2 terms__

You can rearrange the terms in an equation with like terms together. Terms are "like" if they both have the same variable or do not have a variable.

1st arrangement: $4x \quad + 6 \quad + 2x \quad - 3 \quad = 33$
2nd arrangement: $4x \quad + 2x \quad + 6 \quad - 3 \quad = 33$

Look at the equation to answer each question.

4. Write the two terms that have x as a variable.

   __$4x$ and $2x$__

5. What is the sum of these two terms?

   __$6x$__

6. Write the two terms that do not have variables.

   __6 and −3__

7. Find the difference of these two terms to combine them.

   __$6 - 3 = 3$__

8. Rewrite the equation with like terms combined.

   __$6x + 3 = 33$__

17

Puzzles, Twisters & Teasers
Just a Bunch of Hot Air

Decide whether or not each equation is correct. Circle the letter above your answer. Use the letters to solve the riddle.

1. $b + 18 + 3b = 74$ $b = 29$
   U (A)
   correct   incorrect

2. $-3 + 4p - 2p = 1$ $p = 4$
   K (C)
   correct   incorrect

3. $2 - 5(x + 2) = -18$ $x = -4$
   N (R)
   correct   incorrect

4. $6(n - 7) + 40 = 4$ $n = 1$
   (O) Q
   correct   incorrect

5. $5n + 7 - 3n = 19$ $n = 3$
   L (S)
   correct   incorrect

6. $10x - 3 - 2x = 4$ $x = 7$
   G (S)
   correct   incorrect

7. $2.5(a + 4) - 3.2 = 14.3$ $a = 3$
   (W) P
   correct   incorrect

8. $10x - 11 - 4x = 43$ $x = 9$
   (I) M
   correct   incorrect

9. $53 = 12(m - \frac{3}{4}) - 16$ $m = 5$
   J (N)
   correct   incorrect

10. $7n - 1 - 2n = 14$ $n = 3$
    (D) H
    correct   incorrect

What is the angriest kind of wind?

__A__   __C__ __R__ __O__ __S__ __S__ __W__ __I__ __N__ __D__

18

63

Practice A
Solving Equations with Variables on Both Sides

Group the terms with the variables on one side of the equal sign and simplify. Do not solve.

1. $9p = 6p + 21$

$$3p = 21$$

2. $-5t - 14 = 2t$

$$-14 = 7t$$

3. $2k = 18 - 4k$

$$6k = 18$$

4. $-6u + 48 = 6u$

$$48 = 12u$$

5. $7m - 25 = 2m$

$$-25 = -5m$$

6. $\frac{5}{9}a = 6 + \frac{4}{9}a$

$$\frac{1}{9}a = 6$$

Solve. Match each equation with its solution.

7. $8n = 6n - 14$ **A.** $n = 8$

8. $7n + 8 = 11n$ **B.** $n = 9$

9. $2n + 8 = 3n$ **C.** $n = -7$

10. $\frac{3}{4}n + 2 = \frac{1}{4}n + 1$ **D.** $n = 2$

11. $9 + 3n = 4n$ **E.** $n = -8$

12. $\frac{5}{8}n + 5 = \frac{7}{8}n + 7$ **F.** $n = -2$

13. Members of the Campus Roller Rink pay a yearly membership fee of $50 plus $5 each time they go skating. Nonmembers pay $10 each time they go skating. How many times would both a member and a nonmember have to go skating in a year in order to pay the same amount? __10 times__

Holt Mathematics

Practice B
Solving Equations with Variables on Both Sides

Group the terms with the variables on one side of the equal sign and simplify.

1. $10t = 6t + 24$

$$4t = 24$$

2. $-6x - 32 = 2x$

$$-32 = 8x$$

3. $j = 20 - 4j$

$$5j = 20$$

4. $-5d + 40 = 5d$

$$40 = 10d$$

5. $9m - 28 = 2m$

$$-28 = -7m$$

6. $\frac{8}{9}x = 8 + \frac{4}{9}x$

$$\frac{4}{9}x = 8$$

Solve.

7. $8k = 6k - 26$

$$k = -13$$

8. $32 - 5v = 3v + 8$

$$v = 3$$

9. $-12y - 10 = -6y + 14$

$$y = -4$$

10. $\frac{5}{8}a + 6 = \frac{3}{4}a$

$$a = 48$$

11. $\frac{1}{4}n + 10 = \frac{2}{3}n$

$$n = 24$$

12. $20 - \frac{1}{5}d = \frac{3}{10}d + 16$

$$d = 8$$

13. Members of the Lake Shawnee Club pay $40 per summer season plus $7.50 each time they rent a boat. Nonmembers pay $12.50 each time they rent a boat. How many times would both a member and a nonmember have to rent a boat in order to pay the same amount? __8 times__

Holt Mathematics

Practice C
Solving Equations with Variables on Both Sides

Group the terms with variables on one side of the equal sign and simplify.

1. $15t = 17 + 8t - 24$

$$7t = -7$$

2. $6a - 42 = 2a + a$

$$-42 = -3a$$

3. $6k - k = 27 - 4k$

$$9k = 27$$

4. $-5f + 40 = 5f + 30$

$$10 = 10f$$

5. $\frac{9}{10}m - 28 = \frac{7}{10}m + 12$

$$\frac{1}{5}m = 40$$

6. $\frac{8}{15}x = 8 + \frac{2}{15}x - \frac{4}{15}x$

$$\frac{2}{3}x = 8$$

Solve. Check each answer.

7. $18r - 2 - 12r = -6r - 26$

$$r = -2$$

8. $10 + 8c + 15 = -2c - 40$

$$c = -6.5$$

9. $-3w + 6 = 8w + 9 - 2w$

$$w = -\frac{1}{3}$$

10. $0.8a - 6 = -0.6a + 22$

$$a = 20$$

11. $\frac{1}{2}n - 1\frac{5}{12} = \frac{2}{3}n - 2\frac{1}{6}$

$$n = 4\frac{1}{2}$$

12. $\frac{13}{15} - \frac{d}{5} = \frac{5}{6} + \frac{d}{3}$

$$d = \frac{1}{16}$$

13. A golf course charges nonmembers $48 per round plus a golf cart rental fee of $24 per round. A member pays a monthly fee of $384 plus a golf cart rental fee of $8 per round. How many rounds would both a member and a nonmember have to play in order to pay the same amount? __6 rounds__

Holt Mathematics

Reteach
Solving Equations with Variables on Both Sides

Some equations have like terms that are on opposite sides of the equal sign. To solve these equations, first identify the like term with the lesser coefficient.

$2x = 25 - 3x$ ← $2x$ and $-3x$ are like terms.

Since $-3 < 2$, $-3x$ is the term with the lesser coefficient. Move $-3x$ to the opposite side of the equation by adding $3x$ to each side of the equation.

$$2x = 25 - 3x$$
$$2x + 3x = 25 - 3x + 3x$$
$$5x = 25$$
$$\frac{5x}{5} = \frac{25}{5}$$
$$x = 5$$

Complete the steps to solve the equation.

1. $5x + 16 = 9x$

The like terms are __$5x$__ and __$9x$__. The like term with the lesser coefficient is __$5x$__.

$$5x + 16 = 9x$$
$$5x - \underline{5x} + 16 = 9x - \underline{5x}$$
$$16 = \underline{4x}$$
$$\frac{16}{4} = \frac{4x}{4}$$
$$\underline{4} = x$$

2. $3a + 1 = 15 - 4a$

The like terms are __$3a$__ and __$-4a$__. The like term with the lesser coefficient is __$-4a$__.

$$3a + 1 = 15 - 4a$$
$$3a + \underline{4a} + 1 = 15 - 4a + \underline{4a}$$
$$\underline{7a} + 1 = 15$$
$$7a + 1 - \underline{1} = 15 - \underline{1}$$
$$7a = \underline{14}$$
$$\frac{7a}{7} = \frac{14}{7}$$
$$a = \underline{2}$$

Solve.

3. $8d = 7d + 13$

$$d = 13$$

4. $-4p - 16 = 4p$

$$p = -2$$

5. $3n = 25 - 2n$

$$n = 5$$

6. $9c - 6 = 2c + 15$

$$c = 3$$

7. $16 - 6k = 20 - 2k$

$$k = -1$$

8. $2h + 8 = 2 - h$

$$h = -2$$

Holt Mathematics

Holt Mathematics

Challenge
Use the Clues

You can write equations using clues to solve problems.

Example: The sum of two numbers is 36. The difference of the two numbers is 8. Find the numbers.

Let x = one number.

Let y = the other number.

The sum of the two numbers is 36.	$x + y = 36$
The difference of the two numbers is 8.	$x - y = 8$
Solve the system of equations.	$x + y = 36$
	$x - y = 8$

The solution is (22, 14), so the two numbers are 22 and 14.

Solve.

1. The sum of two numbers is 33. The difference of the two numbers is 5. Find the numbers.

   __14 and 19__

2. The sum of two numbers is 85. The difference of the two numbers is 13. Find the numbers.

   __36 and 49__

3. One number is 18 more than another number. The sum of the two numbers is 60. Find the numbers.

   __21 and 39__

4. The sum of two numbers is 99. The difference of the two numbers is 9. Find the numbers.

   __45 and 54__

5. The sum of two numbers is 42. One number is twice the other number. Find the numbers.

   __14 and 28__

6. The sum of two numbers is 43. One number is equal to 5 less than 3 times the other number. Find the numbers.

   __12 and 31__

23   **Holt Mathematics**

Problem Solving
Solving Equations with Variables on Both Sides

Write the correct answer.

1. Five added to twice Erik's age is the same as 3 times his age minus 2. How old is Erik?

   __7 years old__

2. Three times the perimeter of a triangle is the same as 75 decreased by twice the perimeter. What is the perimeter of the triangle?

   __15 units__

3. The area of a pentagon increased by 27 is the same as four times the area of the pentagon, minus 15. What is the area of the pentagon?

   __14 square units__

4. To repair body damage on a car, AutoBody charges $125, plus $18 per hour. CarCare charges $200, plus $12 per hour. Determine the number of hours for which the two body shops will cost the same.

   __12.5 hours__

Choose the letter for the best answer.

5. Sandy and Suzanne are planting flower pots around the school building. Sandy has planted 33 pots and is planting at the rate of 10 pots per hour. Suzanne has planted 25 pots and is planting at the rate of 14 pots per hour. In how many hours will they have planted the same number of flower pots?

   A 3 hr   **C** 2 hr
   B 2.5 hr   **D** 1 hr

6. The length of the sides of a square measure $2x - 5$. The length of a rectangle measures $2x$, and the width measures $x + 2$. For what value of x is the perimeter of the square the same as the perimeter of the rectangle?

   F $x = 2$ **H** $x = 10$
   G $x = 7$ **J** $x = 12$

7. Louisa used Downtown Taxi, which charges $2 for the first mile and $1.10 for each additional mile. Pietro used Uptown Cab, which charges $5 for the first mile and $0.95 for each additional mile. They paid the same amount and traveled the same distance. How far did they travel?

   A 25 mi   **C** 20 mi
   B 21 mi   **D** 15 mi

8. Toni bought some beach towels on sale for $8 each. Theo bought the same number of beach towels at the full price of $12. Toni's total was $24 less than Theo's total. How many beach towels did they each buy?

   F 6 towels   **H** 9 towels
   G 8 towels   **J** 12 towels

24   **Holt Mathematics**

Reading Strategies
Follow a Procedure

To solve equations with variables on both sides, follow these steps.

$8x - 3 = 2x + 9$

Step 1: Get all variable terms on one side of the equation. $8x - 2x - 3 = 2x - 2x + 9$ $6x - 3 = 9$ Subtract $2x$ from both sides.

Step 2: Isolate the variable term. Use the opposite operation. $6x - 3 + 3 = 9 + 3$ $6x = 12$ Add 3 to both sides.

Step 3: Use the inverse operation and calculate. $\frac{6x}{6} = \frac{12}{6}$ $x = 2$ Divide both sides by 6.

Answer the following questions.

1. What is the first step to solve an equation with variables on both sides?

   __Get all the variable terms on one side of the equation.__

2. What operation was performed to get the variables on one side?

   __subtraction__

3. Write the equation that results after the first step is complete.

   $6x - 3 = 9$

4. What is the second step to solve the equation?

   __Get the other values on the opposite side of the equation.__

5. What operation was performed to get 3 on the opposite side?

   __addition__

6. Write the equation the way it looks after the second step is complete.

   $6x = 12$

7. What operation was performed last?

   __division__

25   **Holt Mathematics**

Puzzles, Twisters & Teasers
All Aboard!

Decide whether or not each solution is correct. Circle the letter above your answer. Then use the letters to solve the riddle.

1. $6m = 4m + 12$ $m = 6$
   W correct   **A** incorrect

2. $5n = 4n + 32$ $n = 4$
   M correct   **I** incorrect

3. $4y = 2y + 40$ $y = 10$
   V correct   **D** incorrect

4. $5n = 3n + 26$ $n = 13$
   E correct   **Q** incorrect

5. $8 + 6a = -2a + 24$ $a = 3$
   J correct   **O** incorrect

6. $19 + 7n = -2n + 37$ $n = 2$
   N correct   **W** incorrect

7. $12h = 9h + 84$ $h = 12$
   P correct   **T** incorrect

8. $\frac{5}{9}x + 83 = \frac{4}{9}x + 92$ $x = 81$
   W correct   **C** incorrect

9. $9t = 4t + 120$ $t = 7$
   G correct   **A** incorrect

10. $4.19d = 74.8 + 1.99d$ $d = 34$
    I correct   **B** incorrect

How do you define a "twip"?

A __W I D E__ you take __O N__ a __T W A I__ n

26   **Holt Mathematics**

Choose an inequality for each situation.

| $x > 10$ | $x \geq 10$ | $x < 10$ | $x \leq 10$ |

1. The temperature today will be at least 10°F. $x \geq 10$

2. The temperature tomorrow will be no more than 10°F. $x \leq 10$

3. Yesterday, there was less than 10 inches of snow. $x < 10$

4. Last Wednesday, there was more than 10 inches of snow. $x > 10$

Complete the graph for each inequality.

5. $a > 3$

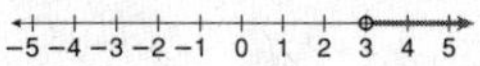

6. $r \leq -2$

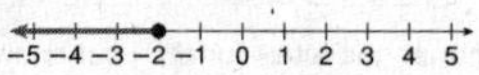

Graph each inequality.

7. $w \geq 0$

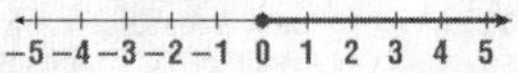

8. $b \leq -4$

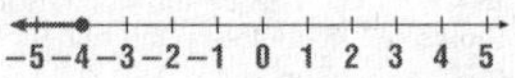

9. $j > 5$

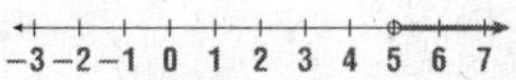

Complete the graph for each compound inequality.

10. $t > 2$ or $t < -1$

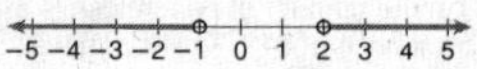

11. $-1 \leq f \leq 4$

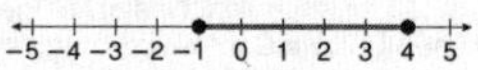

Graph each compound inequality.

12. $y \geq 1$ or $y < -3$

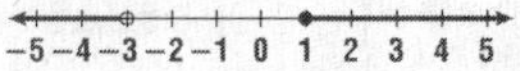

13. $-4 < p < 2$

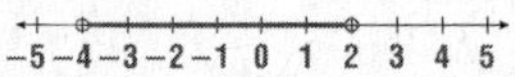

27
Holt Mathematics

Write an inequality for each situation.

1. The temperature today will be at most 50°F. $x \leq 50$

2. The temperature tomorrow will be above 70°F. $x > 70$

3. Yesterday, there was less than 2 inches of rain. $x < 2$

4. Last Monday, there was at least 3 inches of rain. $x \geq 3$

Graph each inequality.

5. $t \leq -2$

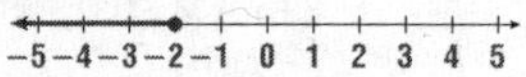

6. $j > -5$

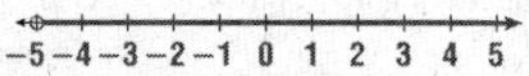

7. $y \leq 0$

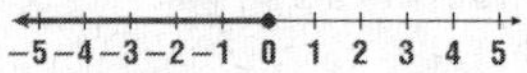

8. $b < \frac{1}{2}$

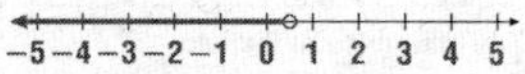

Graph each compound inequality.

9. $f > 3$ or $f < -2$

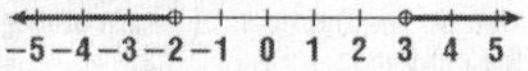

10. $-4 \leq w \leq 4$

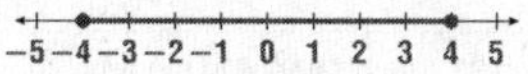

11. $b < 0$ or $b \geq 5$

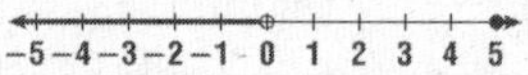

12. $y \geq 3$ or $y \leq -1$

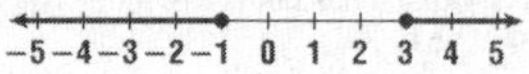

13. $-4 < m < -2$

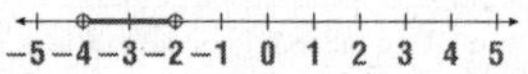

28
Holt Mathematics

Graph each inequality or compound inequality.

1. $x > 14$

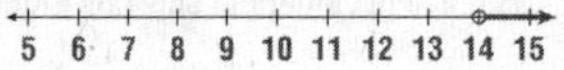

2. $t \leq -22$

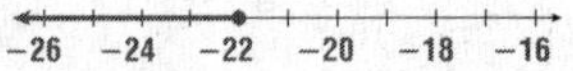

3. $j > -35$

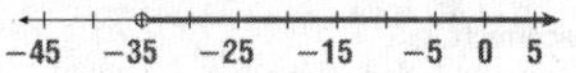

4. $f > 12$ or $f < -12$

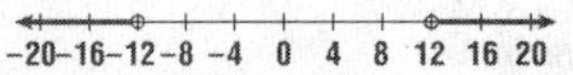

5. $-14 \leq w \leq 2$

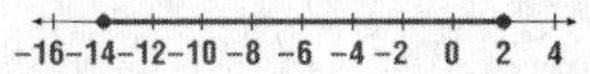

6. $b < 0$ or $b \geq 15$

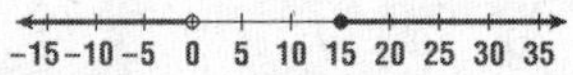

Write each statement using inequality symbols.

7. The number x is between -9 and -19.

$-19 < x < -9$

8. The number y is at most 5.

$y \leq 5$

9. The number g is greater than or equal to 4 or less than -2.

$g \geq 4$ or $g < -2$

10. The number p is less than or equal to -8.

$p \leq -8$

Write an inequality shown by each graph.

11. $x > 3.5$

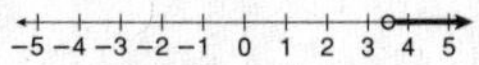

12. $-3 \leq x \leq 1.5$

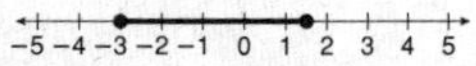

13. $x < -5$ or $x \geq -2$

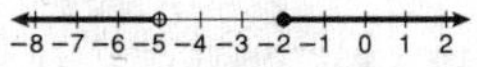

14. $x \leq -2\frac{3}{4}$

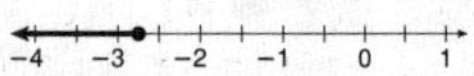

29
Holt Mathematics

An **equation** is a statement that says two quantities are equal. An **inequality** is a statement that says two quantities are **not** equal.

The chart shows symbols and phrases that indicate inequalities.

<	>	≤	≥
Less than	Greater than	Less than or equal to	Greater than or equal to
Fewer than	More than	At most	At least
Below	Above	No more than	No less than

Complete the inequality for each situation.

1. No more than 200 people can be seated in the restaurant.

   number of people seated in restaurant $\leq$ 200

2. The waiting time for a table is at least 20 minutes.

   waiting time $\geq$ 20 minutes

3. The price of all special dinner entrees is below $10.

   special dinner entrees $<$ $10

4. The Yoshida family spent more than $40 for dinner.

   Yoshida family spent $>$ $40

An inequality can be shown on a graph.

The graph shows:	Inequality	Graph
All numbers greater than 3	$x > 3$	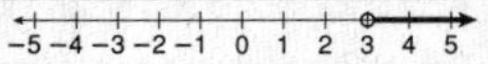 The *open circle* at 3 shows that the value 3 is **not** included in the graph.
All numbers greater than or equal to 3	$x \geq 3$	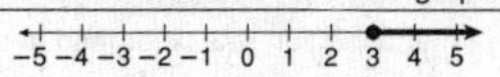 The *closed circle* at 3 shows that the value 3 **is** included in the graph.
All numbers less than 3	$x < 3$	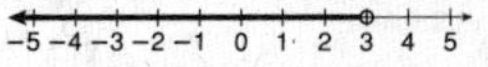
All numbers less than or equal to 3	$x \leq 3$	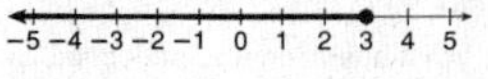

30
Holt Mathematics

66
Holt Mathematics

Reteach
Inequalities (continued)

Graph each inequality.

5. $x > -4$
- Draw an open circle at -4.
- Read $x > -4$ as "x is greater than -4."
- Draw an arrow to the right of -4.

6. $x \leq 1$
- Draw a closed circle at 1.
- Read $x \leq 1$ as "x is less than or equal to 1."
- Draw an arrow to the left of 1.

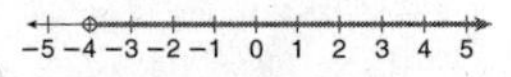 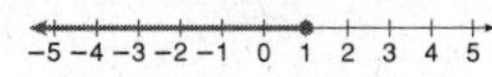

7. $a > -1$

8. $y \leq 3$

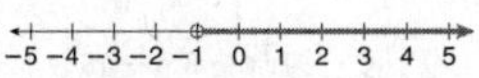 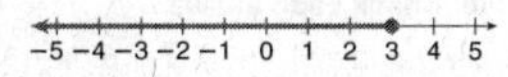

A **compound inequality** is a combination of two inequalities.

The graph shows:	Inequality	Graph
All numbers from -2 to 2	$-2 \leq x \leq 2$	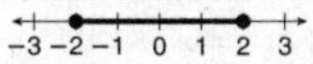 ![graph] $-3\ -2\ -1\ \ 0\ \ 1\ \ 2\ \ 3$
All numbers greater than 2 *or* less than -2	$x > 2$ or $x < -2$	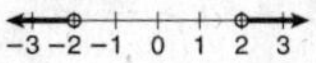 ![graph] $-3\ -2\ -1\ \ 0\ \ 1\ \ 2\ \ 3$

Graph each compound inequality.

9. $0 < x \leq 3$
- Draw an open circle at 0 and a closed circle at 3.
- Shade the line between 0 and 3.

10. $x \geq 4$ or $x \leq -1$
- Draw a closed circle at -1 and a closed circle at 4.
- Shade the line to the left of -1 and to the right of 4.

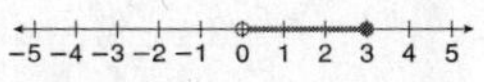 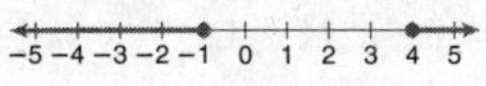

Holt Mathematics

Challenge
Square Equations

Some equations can be solved using square roots.

Example 1:
Solve $x^2 = 25$.
Think: $5^2 = 25$, and $(-5)^2 = 25$.
So, the equation has two solutions.
$x = -5, 5$

Example 2:
Solve $\frac{w}{2} = \frac{8}{w}$.
First, cross multiply: $w^2 = 16$.
Think: $4^2 = 16$, and $(-4)^2 = 16$.
So, the equation has two solutions.
$y = -4, 4$

Solve.

1. $x^2 = 64$
$x = -8, 8$

2. $y^2 = 81$
$y = -9, 9$

3. $n^2 = 121$
$n = -11, 11$

4. $a^2 = 49$
$a = -7, 7$

5. $m^2 = 100$
$m = -10, 10$

6. $s^2 = 169$
$s = -13, 13$

7. $\frac{c}{3} = \frac{3}{c}$
$c = -3, 3$

8. $\frac{k}{4} = \frac{25}{k}$
$k = -10, 10$

9. $\frac{e}{4} = \frac{16}{e}$
$e = -8, 8$

10. $\frac{9}{t} = \frac{t}{4}$
$t = -6, 6$

11. $\frac{16}{d} = \frac{d}{9}$
$d = -12, 12$

12. $z^2 - 1 = 35$
$z = -6, 6$

13. $p^2 - 5 = 20$
$p = -5, 5$

14. $h^2 + 14 = 63$
$h = -7, 7$

15. $29 + v^2 = 110$
$v = -9, 9$

Holt Mathematics

Problem Solving
Inequalities

Write the correct answer.

The American College of Sports Medicine recommends exercising at an intensity of 60% to 90% of your maximum heart rate.

	Heart Rates by Age	
Age	Maximum Heart Rate	Target Range
20–24	200	120–180
25–29	195	117–176
30–34	190	114–171
35–39	185	111–167
40–44	180	108–162

1. Mara is 25 years old. Write a compound inequality to represent her target heart rate range while bike riding.
$117 \leq x \leq 176$

2. Leia is 38 years old. Write a compound inequality to represent the zone between her maximum heart rate and the upper end of her target range.
$167 < x \leq 185$

3. Rudy is 42 years old. Write an inequality to represent at least 70% of his maximum heart rate.
$x \geq 0.7 \cdot 180; \ x \geq 126$

4. Write a compound inequality to represent 60% to 90% of your maximum heart rate in ten years.
Possible answer: $120 \leq x \leq 180$

Choose the letter for the graph that represents each statement.

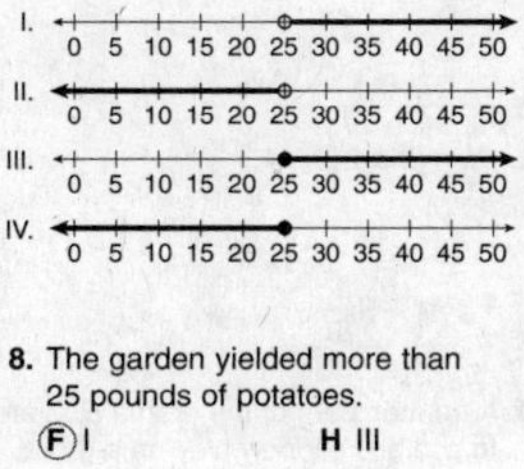

5. Alena decided to pay not more than $25 to get her old bike repaired.
- A I
- B II
- C III
- (D) IV

6. It was so cold last week that the temperature never reached 25°F.
- F I
- (G) II
- H III
- J IV

I. ![number line] 0 5 10 15 20 25 30 35 40 45 50
II. ![number line] 0 5 10 15 20 25 30 35 40 45 50
III. ![number line] 0 5 10 15 20 25 30 35 40 45 50
IV. ![number line] 0 5 10 15 20 25 30 35 40 45 50

7. There were at least 25 people ahead of Ivan in the cafeteria line.
- A I
- B II
- (C) III
- D IV

8. The garden yielded more than 25 pounds of potatoes.
- (F) I
- G II
- H III
- J IV

Holt Mathematics

Reading Strategies
Connecting Words and Symbols

An **inequality** is a comparison of two unequal values. This chart will help you understand both words and symbols for inequalities.

The team has scored fewer than 5 runs in each game. "Fewer than 5" means "less than 5." Symbol for "less than 5": < 5	No more than 8 people can ride in the elevator. "No more than 8" Means "**8 or less than 8.**" Symbol for "less than or equal to 8": ≤ 8

Inequalities

More than 25 students try out for the team each year. "More than 25" means "**a number greater than 25.**" Symbol for "greater than 25": > 25	There are at least 75 fans at each home game. "At least 75" means "75 or more" or "**a number greater than or equal to 75.**" Symbol for "greater than or equal to 75": ≥ 75

Use the chart to answer each question.

1. What is an inequality?
a comparison of two unequal values

2. Explain the difference between the symbols $<$ and $\leq$.
The symbol $<$ means less than and the symbol $\leq$ means less than or equal to.

3. Explain the difference between the symbols $>$ and $\geq$.
The symbol $>$ means greater than and the symbol $\geq$ means greater than or equal to.

4. Write $<$, $>$, $\leq$ or $\geq$ to describe the number of students in each homeroom: There is a limit of 30 students for each homeroom.
≤ 30 students

Holt Mathematics

Holt Mathematics

Puzzles, Twisters & Teasers
Hard to Swallow!

Find and circle words from the list in the word search (horizontally, vertically or diagonally). Find a word that answers the riddle. Circle it and write it on the line.

inequality   compound   value   symbol   graph
less   greater   equal   more   fewer

```
I N E Q U A L I T Y V F W
L Z X C V M N M K I A E N
E A S D F G O J K L L W J
S W E R K C F R T G U E K
S Y M B O L M K E U E R U
V F R G R A P H B G T N O
Y H N J E Q U A L C D E P
H J G R E A T E R A S E C
A S D F C O M P O U N D S
L I M E S T O N E N M J V
Q W E R T Y U I O P L K X
```

What kind of stone is the most sour? _______ limestone _______

    35   **Holt Mathematics**

Practice A
Solving Inequalities by Adding or Subtracting

Solve. Then match each solution set with its graph.

1. $r - 1 > 2$ ___ $r > 3$

2. $m + 3 \le 6$ ___ $m \le 3$

3. $x - 4 < -1$ ___ $x < 3$

4. $k + 2 \ge 5$ ___ $k \ge 3$

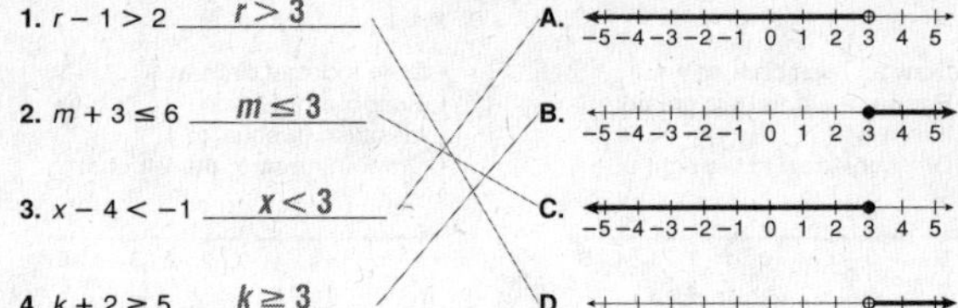

A. $-5\ -4\ -3\ -2\ -1\ \ 0\ \ 1\ \ 2\ \ 3\ \ 4\ \ 5$

B. $-5\ -4\ -3\ -2\ -1\ \ 0\ \ 1\ \ 2\ \ 3\ \ 4\ \ 5$

C. $-5\ -4\ -3\ -2\ -1\ \ 0\ \ 1\ \ 2\ \ 3\ \ 4\ \ 5$

D. $-5\ -4\ -3\ -2\ -1\ \ 0\ \ 1\ \ 2\ \ 3\ \ 4\ \ 5$

Solve. Check each answer.

5. $a + 7 > 2$ 6. $h - 9 \le 3$ 7. $v - 8 < -16$

___ $a > -5$ ___   ___ $h \le 12$ ___   ___ $v < -8$ ___

8. $14 + y > -7$ 9. $j - 17 \ge -8$ 10. $t + 25 \le 9$

___ $y > -21$ ___   ___ $j \ge 9$ ___   ___ $t \le -16$ ___

11. $w + 23 < -35$ 12. $f + 39 \ge 11$ 13. $54 + b \le 45$

___ $w < -58$ ___   ___ $f \ge -28$ ___   ___ $b \le -9$ ___

14. Friday's high temperature was 20°F. Saturday's high temperature was forecast to be no more than 8°F warmer than Friday's high temperature. According to the forecast, what are the possible high temperatures for Saturday?

The possible high temperature is at most 28⁰F.

    36   **Holt Mathematics**

Practice B
Solving Inequalities by Adding or Subtracting

Solve. Then graph each solution set on a number line.

1. $y - 5 > -2$ ___ $y > 3$ $-5\ -4\ -3\ -2\ -1\ \ 0\ \ 1\ \ 2\ \ 3\ \ 4\ \ 5$

2. $n + 5 \le 11$ ___ $n \le 6$ $-2\ -1\ \ 0\ \ 1\ \ 2\ \ 3\ \ 4\ \ 5\ \ 6\ \ 7\ \ 8$

3. $x + 4 < -1$ ___ $x < -5$ $-9\ -8\ -7\ -6\ -5\ -4\ -3\ -2\ -1\ \ 0\ \ 1$

4. $h + 20 > 2$ ___ $h > -18$ $-20\ \ -18\ \ -16\ \ -14\ \ -12\ \ -10$

5. $p + 9 \ge -3$ ___ $p \ge -12$ $-18\ \ -16\ \ -14\ \ -12\ \ -10\ \ -8$

6. $s - 7 < -16$ ___ $s < -9$ $-10\ -9\ -8\ -7\ -6\ -5\ -4\ -3\ -2\ -1\ \ 0$

Solve. Check each answer.

7. $41 + g > 27$ 8. $w + 23 \ge -18$ 9. $a + 15 \le 9$

___ $g > -14$ ___   ___ $w \ge -41$ ___   ___ $a \le -6$ ___

10. $z + 27 < 16$ 11. $-3 \le t + 17$ 12. $78 \ge b + 64$

___ $z < -11$ ___   ___ $-20 \le t$ ___ $14 \ge b$

13. In order for a field trip to be scheduled, at least 30 students must sign up. So far, 23 students have signed up. At least how many more students must sign up in order for the field trip to be scheduled?

At least 7 more students must sign up.

    37   **Holt Mathematics**

Practice C
Solving Inequalities by Adding or Subtracting

Solve. Then graph each solution set on a number line.

1. $a - (-5) > 2$ ___ $a > -3$ $-5\ -4\ -3\ -2\ -1\ \ 0\ \ 1\ \ 2\ \ 3\ \ 4\ \ 5$

2. $m - 3 \le 7 - 11$ ___ $m \le -1$ $-5\ -4\ -3\ -2\ -1\ \ 0\ \ 1\ \ 2\ \ 3\ \ 4\ \ 5$

3. $x + 4 > -1 + 8$ ___ $x > 3$ $-5\ -4\ -3\ -2\ -1\ \ 0\ \ 1\ \ 2\ \ 3\ \ 4\ \ 5$

Solve. Check each answer.

4. $j - 14 + 20 > 25$ 5. $w - 9 - (-1) \ge 3$ 6. $r - 17 < 14 - 16$

___ $j > 19$ ___   ___ $w \ge 11$ ___   ___ $r < 15$ ___

7. $24 - 11 + k > 27 - 30$ 8. $x - (-13) + 18 \le 8 - (-9)$ 9. $9 - 15 \le b + 16$

___ $k > -16$ ___   ___ $x \le -14$ ___   ___ $b \ge -22$ ___

10. $t - 5.3 + 2.8 < 9.5 - 11.2$ 11. $s + 6.2 - 5.9 \ge 1.3 + 3.7$

___ $t < 0.8$ ___   ___ $s \ge 4.7$ ___

12. $1\frac{1}{2} + z \le 2\frac{7}{8} - 4\frac{3}{4}$ 13. $n + \frac{1}{6} > 1\frac{1}{3} - 2$

___ $z \le -3\frac{3}{8}$ ___   ___ $n > -\frac{5}{6}$ ___

14. Nine members of the drama club are going to New York to see a Broadway play as a group. The total cost for the tickets, including a $15.00 handling fee, will be no more than $285.00. What is the maximum cost of each ticket?   **$30.00**

    38   **Holt Mathematics**

Inequalities with variables may have more than one solution. All the solutions of an inequality are called the **solution set**.

You can solve an inequality involving addition or subtraction just as you would solve an equation.

Complete the steps to solve, graph, and check the inequality.

1.
$$n - 13 < 7$$
$$n - 13 + \underline{13} < 7 + \underline{13}$$
$$n < \underline{20}$$

Graph the inequality.

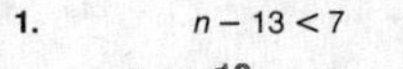

Check:

Pick any number in the solution set of $n < 20$.
$19 < 20$

Substitute 19 into the inequality.

$$n - 13 < 7$$
$$\underline{19} - 13 \overset{?}{<} 7$$
$$\underline{6} \overset{?}{<} 7 ✔$$

2.
$$x + 18 \geq 2$$
$$x + 18 - \underline{18} \geq 2 - \underline{18}$$
$$x \geq \underline{-16}$$

Graph the inequality.

Check:

Pick any number in the solution set of $x \geq -16$.
$0 \geq -16$

Substitute 0 into the inequality.

$$x + 18 \geq -2$$
$$\underline{0} + 18 \overset{?}{\geq} -2$$
$$\underline{18} \overset{?}{\geq} -2 ✔$$

Solve. Then graph each solution set.

3. $d + 3 > -5$ $\underline{d > -8}$

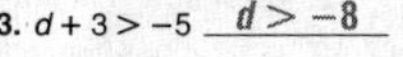

4. $s - 10 < -6$ $\underline{s < 4}$

Solve. Check each answer.

5. $y + 9 > 20$

$\underline{y > 11}$

6. $h + 17 \leq -6$

$\underline{h \leq -23}$

7. $a - 4 \geq -18$

$\underline{a \geq -14}$

39   **Holt Mathematics**

The absolute value of a number x is:

$$|x| = x \text{ if } x > 0 \qquad \text{and} \qquad |x| = -x \text{ if } x < 0$$

When you solve an absolute value equation, you must consider both of these cases.

Example: Solve $|n + 8| = 20$.

Case 1: $n + 8 = 20$
    $n = 12$

Case 2: $-(n + 8) = 20$
    $-n - 8 = 20$
    $-n = 28$
    $n = -28$

The equation has two solutions: $n = 12, -28$

Solve.

1. $|t - 4| = 9$

$\underline{t = -5, 13}$

2. $|6s| = 72$

$\underline{s = -12, 12}$

3. $|y| + 3 = 7$

$\underline{y = -4, 4}$

4. $|g + 6| = 3$

$\underline{g = -9, -3}$

5. $|2m - 7| = 29$

$\underline{m = -11, 18}$

6. $|4p - 1| = 15$

$\underline{p = -\dfrac{7}{2}, 4}$

7. $|c + 7| = 13$

$\underline{c = -20, 6}$

8. $|3n - 4| = 11$

$\underline{n = -\dfrac{7}{3}, 5}$

9. $|2e| - 8 = 24$

$\underline{e = -16, 16}$

10. $|5x| - 8 = 32$

$\underline{x = -8, 8}$

11. $|9k| + 3 = 48$

$\underline{k = -5, 5}$

12. $|2s - 6| = 20$

$\underline{s = -7, 13}$

13. $|3w - 5| = 4$

$\underline{w = \dfrac{1}{3}, 3}$

14. $|4b - 7| = 1$

$\underline{b = \dfrac{3}{2}, 2}$

15. $|6r - 1| = 17$

$\underline{r = -\dfrac{8}{3}, 3}$

40   **Holt Mathematics**

Write the correct answer.

1. A small car averages up to 29 more miles per gallon of gas than an SUV. If a small car averages 44 miles per gallon, what is the average miles per gallon for an SUV?

$\underline{\text{at least 15 mi per gal}}$

2. Carlos is taking a car trip that is more than 240 miles, depending on the route he chooses. He has already driven 135 miles. How much farther does he have to go?

$\underline{\text{more than 105 miles}}$

3. Driving into the city usually takes 25 minutes. If there is a lot of traffic, the trip can take up to 45 minutes. How much additional time should you allow during a heavy traffic period?

$\underline{\text{no more than 20 min}}$

4. To qualify for the heavyweight wrestling division, Kobe must weigh at least 180 pounds. If Kobe weighs 168 pounds now, how much weight should he gain?

$\underline{\text{at least 12 lb}}$

Choose the letter for the best answer.

5. On one day, the range of temperatures in one state was at most 27°. If the lowest temperature in the state was 59°, what was the highest temperature?

A $t > 86°$ C $t = 86°$
Ⓑ $t \leq 86°$ D $t \geq 86°$

6. The highest possible score on the Scholastic Aptitude Test is 2,400. Rebecca scored 1780. She needs a score of at least 1,950 to qualify for a scholarship. How much higher must her score be?

F $s \leq 170$ Ⓗ $s \geq 170$
G $s \leq 450$ J $s \geq 450$

7. Romero is saving to buy an Apex Model 12 Computer. The lowest price that Romero can find for the computer is $1,250. Romero now has $825. His grandmother is going to give him another $200. How much more money does Romero need?

A $x < \$225$ Ⓒ $x \geq \$225$
B $x < \$425$ D $x \geq \$425$

8. The seating capacity of the school gym is 550. So far, there are 210 fans at a basketball game. How many more fans could attend the game?

F $f \geq 340$
G $f > 210$
Ⓗ $f \leq 340$
J $f < 340$

41   **Holt Mathematics**

You can use this chart to help you understand solving two-step inequalities.

Solution of the Inequality	Solution Set
Any value of the variable that makes the inequality true.	All possible solutions of an inequality.

Solving Inequalities by Adding or Subtracting

Addition Inequality
$x + 4 > 25$
$x + 4 - 4 > 25 - 4$
$x > 21$

Subtraction Inequality
$y - 8 < 12$
$y - 8 + 8 < 12 + 8$
$y < 20$

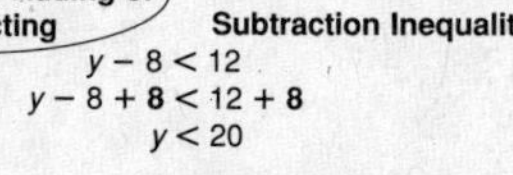

Use the graphic organizer to answer the following questions.

1. What is a solution of an inequality?

$\underline{\text{any value for the variable that makes the inequality true}}$

2. What is the solution set?

$\underline{\text{all the possible solutions of an inequality}}$

3. What are 3 solutions for $x > 21$?

$\underline{\text{possible answers: 22, 23, 24}}$

4. What are 3 solutions for $y < 20$?

$\underline{\text{possible answers: 19, 18, 17}}$

5. Why is $x + 4 > 25$ called an Addition Inequality?

$\underline{\text{4 is added to the variable } x \text{ in the inequality.}}$

42   **Holt Mathematics**

Puzzles, Twisters & Teasers
Are All Puzzles Created Equal?

Across

1. You can use an ____ to solve a problem.

4. Inequalities with ____ usually have many solutions.

5. Any value of a variable that makes a ____ true is a solution of the inequality.

7. When ____ your solution, you should choose a number in the solution set that is easy to work with.

9. All of the solutions of an inequality are called the ____ set.

Down

2. You can find solution sets for inequalities by ____ the variable.

3. You can check your answer by choosing any number in the solution set and ____ it into the original inequality.

6. You can solve one-____ inequalities by adding or subtracting.

8. When you add or ____ the same number to both sides of an inequality, the statement will still be true.

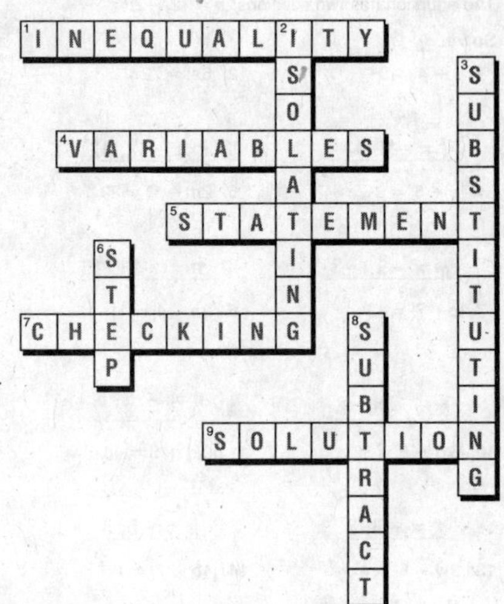

43   **Holt Mathematics**

Practice A
Solving Inequalities by Multiplying or Dividing

Solve. Choose the letter for the best answer.

1. $5n \le -20$

    (A) $n \le -4$ C $n \ge -4$
    B $n < -4$ D $n \ge 4$

2. $-3k > -21$

    F $k > 7$ (H) $k < 7$
    G $k > -7$ J $k < -7$

3. $\frac{m}{5} \ge -1$

    A $m \le -5$ C $m \le 5$
    (B) $m \ge -5$ D $m \ge 5$

4. $\frac{v}{-8} < -4$

    F $v \ge 32$ H $v < 32$
    G $v \le -32$ (J) $v > 32$

5. $-2y < 12$

    A $y < 6$ C $y > 6$
    B $y < -6$ (D) $y > -6$

6. $\frac{h}{5} \le 4.2$

    (F) $h \le 21$ H $h \ge 21$
    G $h \le 22$ J $h \ge 22$

7. $\frac{z}{-9} \le -4$ $z \ge 36$

8. $\frac{b}{3} > 15$ $b > 45$

9. $\frac{m}{6} < -2.5$ $m < -15$

Solve. Check each answer.

10. $7c < -42$ $c < -6$

11. $-20a \ge -100$ $a \le 5$

12. $-8t > 5$ $t < -\frac{5}{8}$

13. It cost \$730 to put on the school play. How many tickets must be sold at \$6 apiece in order to make a profit?

    at least 122 tickets must be sold.

44   **Holt Mathematics**

Practice B
Solving Inequalities by Multiplying or Dividing

Solve.

1. $\frac{n}{5} \le 1.6$ $n \le 8$

2. $\frac{b}{3} > -8$ $b > -24$

3. $\frac{a}{3} \ge -9$ $a \ge -27$

4. $\frac{t}{-6} < -7$ $t > 42$

5. $\frac{s}{-12} \le -5$ $s \ge 60$

6. $\frac{r}{5.3} \le 6$ $r \le 31.8$

Solve. Check each answer.

7. $8c < -64$ $c < -8$

8. $-16a \ge -24$ $a \le 1\frac{1}{2}$

9. $-12t > 9$ $t < -\frac{3}{4}$

10. $-3s \le -180$ $s \ge 60$

11. $18b > -24$ $b > -1\frac{1}{3}$

12. $-6m \ge 4$ $m \le -\frac{2}{3}$

13. It cost Sophia \$530 to make wind chimes. How many wind chimes must she sell at \$12 apiece to make a profit?

    at least 45 wind chimes

14. It cost the Wilson children \$55 to make lemonade. How many glasses must they sell at 75¢ each to make a profit?

    at least 74 glasses

15. Jorge's soccer team is having its annual fund raiser. The team hopes to earn at least three times as much as it did last year. Last year the team earned \$87. What is the team's goal for this year?

    at least \$261

45   **Holt Mathematics**

Practice C
Solving Inequalities by Multiplying or Dividing

Solve.

1. $14w \le -2.8$ $w \le -0.2$

2. $-0.6v > -54$ $v < 90$

3. $\frac{u}{7} \ge -9.8$ $u \ge -68.6$

4. $\frac{f}{-0.5} < -0.2$ $f > 0.1$

5. $-2.5e < 1.5$ $e > -0.6$

6. $\frac{m}{4.5} \le 6.3$ $m \le 28.35$

Solve. Check each answer.

7. $4p < (-2)^3$ $p < -2$

8. $-8c \ge -4 \cdot 6$ $c \le 3$

9. $-12a > 8^2$ $a < -5\frac{1}{3}$

10. $\frac{x}{-4 \cdot 3} \le -5 \cdot (-3)$ $x \ge 180$

11. $\frac{d}{6} > -3\frac{2}{3}$ $d > -22$

12. $1\frac{1}{6}k < 14$ $k < 12$

Write an inequality, then solve.

13. Four families are planning to make a joint purchase at a warehouse club. In order to purchase no more than 15 pounds of potatoes for each family, how many potatoes should they buy?

    $\frac{x}{4} \le 15$, $x \le 60$ lb

14. Lila's swim team is raising money to help buy a new pool cover. The team has raised \$635. This is 65% of their goal. At least how much more do they need to reach their goal?

    $x \ge \$341.92$, they need at least \$341.92

46   **Holt Mathematics**

When you multiply or divide both sides of an inequality by the same *positive* number, the direction of the inequality symbol remains the same.

$3p < -18$

Divide both sides by *positive* 3.

$\frac{3p}{3} < \frac{-18}{3}$

$p < -6$

$\frac{x}{5} \geq -4$

Multiply both sides by *positive* 5.

$5 \cdot \frac{x}{5} \geq 5 \cdot -4$

$x \geq -20$

When you multiply or divide both sides of an inequality by the same *negative* number, the direction of the inequality symbol is reversed.

$-6y > -42$

Divide both sides by *negative* 6.

$\frac{-6y}{-6} < \frac{-42}{-6}$

$y < 7$

$\frac{m}{-2} \leq 15$

Multiply both sides by *negative* 2.

$-2 \cdot \frac{m}{-2} \geq -2 \cdot 15$

$m \geq -30$

Solve.

1. $\frac{a}{8} < -3$

Multiply by $\underline{8}$. The direction of the inequality symbol

$\underline{\text{remains the same}}$.

$8 \cdot \frac{a}{8} \boxed{<} 8 \cdot -3$

$a \boxed{<} -24$

2. $-4s \geq -36$

Divide by $\underline{-4}$. The direction of the inequality symbol

$\underline{\text{is reversed}}$.

$\frac{-4s}{-4} \boxed{\leq} \frac{-36}{-4}$

$s \boxed{\leq} 9$

Solve. Check each answer.

3. $\frac{r}{-7} \geq 2$

$r \leq -14$

4. $9b < -54$

$b < -6$

5. $-3n > -36$

$n < 12$

6. $\frac{c}{6} < -8$

$c < -48$

47

Holt Mathematics

Use the definition of absolute value to solve absolute value inequalities.

Remember: $|x| = x$ if $x > 0$ and $|x| = -x$ if $x < 0$

Example 1: $|10 - 2x| > 6$

Case 1: $10 - 2x > 6$

$-2x > -4$

$x < 2$

Case 2: $-(10 - 2x) > 6$

$-10 + 2x > 6$

$2x > 16$

$x > 8$

Solution: $x < 2$ or $x > 8$

Example 2: $|2y - 3| \leq 7$

Case 1: $2y - 3 \leq 7$

$2y \leq 10$

$y \leq 5$

Case 2: $-(2y - 3) \leq 7$

$-2y + 3 \leq 7$

$-2y \leq 4$

$y \geq -2$

Solution: $-2 \leq y \leq 5$

Solve.

1. $|m - 6| \leq 8$

$-2 \leq m \leq 14$

2. $|x + 4| < 10$

$-14 < x < 6$

3. $|6 - y| \leq 2$

$4 \leq y \leq 8$

4. $|b + 5| > 8$

$b < -13$ or $b > 3$

5. $|4 - k| \geq 5$

$k \leq -1$ or $k \geq 9$

6. $|2 - p| - 3 < 1$

$-2 < p < 6$

7. $4 - 3|e| < 1$

$e < -1$ or $e > 1$

8. $\left|\frac{h}{6}\right| \geq 1$

$h \leq -6$ or $h \geq 6$

9. $\left|\frac{n}{4}\right| \leq 2$

$-8 \leq n \leq 8$

10. $|2a + 1| \geq 1$

$a \leq -1$ or $a \geq 0$

11. $|z - 2| < 5$

$-3 < z < 7$

12. $\left|\frac{c}{2} - 3\right| \leq 2$

$2 \leq c \leq 10$

48

Holt Mathematics

Write the correct answer.

1. U.S. Postal Service regulations state that a package can be mailed using Parcel Post rates if it weighs no more than 70 pounds. What is the maximum number of books weighing 12 pounds each that can be mailed in one box using Parcel Post rates?

at most 5 books

2. Marc wants to buy a set of at least 6 antique chairs for his dining room. He has decided to spend no more than $390. What is the most he can spend per chair?

no more than $65

3. Alfonso earns $9.00 per hour working part-time as a lab technician. He wants to earn more than $144 this week. At least how many hours does Alfonso have to work?

more than 16 hours

4. Mrs. Menendez invited 8 children to her son's birthday party. She wants to make sure that each child gets at least 4 small prizes. At least how many prizes should she buy?

at least 32 prizes

Choose the letter for the best answer.

5. The Computer Club spent $2,565 on mouse pads. The members plan to sell the mouse pads during the book fair. If they charge $9.50 for each mouse pad, how many must they sell in order to make a profit?

A $n \leq 271$

B $n \leq 270$

C $n \geq 271$

D $n \geq 270$

6. The Parents Organization bought 1,000 bumper stickers at $1.25 each to sell at football games. They want to make at least $750 profit. What should be the selling price of the bumper stickers?

F $p \geq \$1.25$

G $p \geq \$1.50$

H $p \geq \$1.75$

J $p \geq \$2.00$

7. In 2000, the national ratio of students to computers with Internet access in public schools was 7:1. Winston School had 623 students. If the school had a lower ratio, how many computers with Internet access did Winston School have?

A $c \geq 90$

B $c \leq 90$

C $c \geq 91$

D $c \leq 91$

8. A new theme park averaged fewer than 2,000 visitors per week during the winter months. What was the average daily attendance?

F $a \geq 290$

G $a \leq 186$

H $a > 385$

J $a < 286$

49

Holt Mathematics

Look at the steps for solving multiplication and division inequalities.

Steps for Solving an Inequality	Division Inequality	Multiplication Inequality
Step 1: Isolate the variable.	$\frac{x}{8} < 5$	$4y > 12$
Step 2: Multiply both sides of a division equation by the same number. Divide both sides of a multiplication equation by the same number.	$\frac{x}{8} \cdot 8 < 5 \cdot 8$	$\frac{4y}{4} > \frac{12}{4}$
Step 3: Compute to solve.	$x < 40$	$y > 3$

Use the chart to answer the following questions.

1. What do you do to get the variable by itself in a division inequality?

multiply

2. What was $\frac{x}{8}$ multiplied by in the example?

8

3. What is the solution of the division inequality?

$x < 40$

4. What are three numbers that make the division inequality true?

Possible answers: 37, 38, 39

5. What do you do to get the variable by itself in a multiplication inequality?

divide

6. What was $4y$ divided by in the multiplication inequality?

4

7. What is the solution to the multiplication inequality?

$y > 3$

8. What are three numbers that are part of the solution?

Possible answers: 4, 5, 6

50

Holt Mathematics

Puzzles, Twisters & Teasers
Bone Tired!

You dog has lost his bone and wants to find it before he takes a nap. Circle the correct symbol to solve each inequality. Then use the number under the correct answer to move through the maze. The first one is done for you.

1. $5w$ __ 517 $w = 111$ (>) <   Move __2__ spaces right
   $555 > 517$ 2   5

2. $3m$ __ -15 $m = 2$ (>) <   Move __8__ spaces down
   $6 > -15$ 8   3

3. $-12y$ __ -60 $y = 3$ (>) <   Move __9__ spaces right
   $-36 > -60$ 9   3

4. $8y$ __ 16 $y = 1$ > (<)   Move __6__ spaces up
   $8 < 16$ 8   6

5. $4x$ __ 9 $x = 5$ (>) <   Move __3__ spaces right
   $20 > 9$ 3   7

6. $25c$ __ 200 $c = 4$ > (<)   Move __2__ spaces up
   $100 < 200$ 4   2

7. $-6r$ __ -48 $r = 12$ > (<)   Move __5__ spaces right
   $-72 < -48$ 9   5

8. $\frac{x}{11}$ __ 3 $x = 22$ > (<)   Move __8__ spaces down
   $\frac{22}{11} < 3$ 3   8

51   **Holt Mathematics**

Practice A
Solving Two-Step Inequalities

Solve. Cross out each inequality in the box that matches a solution. Then graph each solution set.

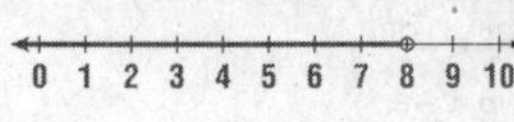

$x > 8$	$x < -8$	$x \geq -8$	$x < 8$	$x \geq 8$	$x \leq -8$

1. $3x - 5 < 19$ ___ $x < 8$

2. $-2x + 12 < -4$ ___ $x > 8$

3. $\frac{x}{4} + 7 \geq 9$ ___ $x \geq 8$

4. $\frac{x}{-2} - 3 \geq 1$ ___ $x \leq -8$

Solve. Then graph each solution set.

5. $7y - 8 > 6$ ___ $y > 2$

6. $-4d + 15 \leq -1$ ___ $d \geq 4$

7. $\frac{r}{-6} + 5 < 7$ ___ $r > -12$

8. Margie has $100. She wants to buy a book for $20 and some CDs for $15 each. At most, how many CDs can Margie buy?

   At most, Margie can buy 5 CD's.

52   **Holt Mathematics**

Practice B
Solving Two-Step Inequalities

Solve. Then graph each solution set on a number line.

1. $5x - 8 < 17$ ___ $x < 5$

2. $\frac{r}{3} + 5 \geq 9$ ___ $r \geq 12$

3. $-4n + 8 < -4$ ___ $n > 3$

4. $\frac{z}{7} - 6 \geq -5$ ___ $z \geq 7$

5. $\frac{w}{-5} + 4 < 9$ ___ $w > -25$

6. $\frac{u}{2} - 5 \leq -9$ ___ $u \leq -8$

Solve.

7. $-7d + 8 > 29$ 8. $4g - 18 \leq -2$ 9. $12 - 3b < 9$

   $d < -3$ $g \leq 4$ $b > 1$

10. $\frac{a}{-4} - 7 < -2$ 11. $9 + \frac{c}{6} \leq 17$ 12. $-\frac{2}{3}p - 8 \geq 4$

   $a > -20$ $c \leq 48$ $p \leq -18$

13. Fifty students in the seventh grade are trying to raise at least $2,000 for sports supplies. They have already raised $750. How much should each student raise, on average, in order to meet the goal?

   On average, each student should raise at least $25.

53   **Holt Mathematics**

Practice C
Solving Two-Step Inequalities

Solve. Then graph each solution set on a number line.

1. $5y + 1 - 8 < 13$ ___ $y < 4$

2. $\frac{r}{8} - 5 - 6 \geq -9$ ___ $r \geq 16$

3. $-7m - 6 < -4 - 9$ ___ $m > 1$

Solve.

4. $\frac{s}{5} - 6 + 3 \geq -5$ 5. $\frac{w}{-2} + 15 < 9 - (-3)$ 6. $\frac{u}{3} - 0.5 \leq -9.5$

   $s \geq -10$ $w > 6$ $u \leq -27$

7. $-4d + 6 - d \leq 21$ 8. $6p - 18 - 2p \leq -2$ 9. $9h - 20 - 3h < -8$

   $d \geq -3$ $p \leq 4$ $h < 2$

10. $\frac{a}{2} - 7\frac{1}{2} + \frac{a}{2} < -2\frac{1}{2}$ 11. $4.5 + \frac{c}{2.3} \leq 6.1$ 12. $-\frac{2}{3}q - \frac{3}{4} \geq \frac{1}{12}$

   $a < 5$ $c \leq 3.68$ $q \leq -1\frac{1}{4}$

13. Max earns $1,800 per month, plus a commission of 3% of his sales. He wants to earn at least $3,000 this month. What amount of sales does Max need this month?

   Max needs at least $40,000 in sales this month.

54   **Holt Mathematics**

---

72   **Holt Mathematics**

Reteach
Solving Two-Step Inequalities

Two-step inequalities can be solved by first undoing addition or subtraction, then undoing multiplication or division.

> Remember to reverse the inequality symbol if you multiply or divide by a negative number.

Complete the steps to solve the inequality. Then graph the solution set.

1.
$$-3x + 35 > -10$$
$$-3x + 35 \underline{-35} > -10 \underline{-35}$$ ← First undo addition or subtraction.
$$-3x > \underline{-45}$$ ← Then undo multiplication or division.
$$\frac{-3x}{-3} \boxed{<} \frac{-45}{-3}$$ ← Divide by −3. The inequality symbol is reversed.
$$x \boxed{<} \underline{15}$$

Graph the inequality.

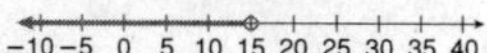

−10 −5 0 5 10 15 20 25 30 35 40

Solve. Then graph each solution set.

2. $7x + 32 < 18$ $\quad \underline{x < -2}$

−5 −4 −3 −2 −1 0 1 2 3 4 5

3. $\frac{t}{4} - 8 \le -5$ $\quad \underline{t \le 12}$

0 2 4 6 8 10 12 14 16 18 20

4. $6f - 6 > 48$ $\quad \underline{f > 9}$

2 3 4 5 6 7 8 9 10 11 12

5. $-4w - 13 \ge 15$ $\quad \underline{w \le -7}$

−9 −8 −7 −6 −5 −4 −3 −2 −1 0 1

6. $\frac{k}{-5} - 6 < -9$ $\quad \underline{k > 15}$

−10 −5 0 5 10 15 20 25 30 35 40

7. $7 - 2p > -5$ $\quad \underline{p < 6}$

−10 −8 −6 −4 −2 0 2 4 6 8 10

55   **Holt Mathematics**

Challenge
Inequalities With Two Variables

You can determine if an ordered pair is a solution of an inequality by substituting the values into the inequality.

- If the inequality is true, the ordered pair is a solution of the inequality.
- If the inequality is false, the ordered pair is not a solution of the inequality.

Example: Is (1, 1) or (2, −1) a solution of the inequality $y < 3x - 2$?

(1, 1)
$y < 3x - 2$
$1 < 3(1) - 2$
$1 < 1$ False
(1, 1) is not a solution.

(2, −1)
$y < 3x - 2$
$-1 < 3(2) - 2$
$-1 < 4$ True
(2, −1) is a solution.

Solve.

1. Which of the following ordered pairs are solutions of $y < 4x - 3$:
$(-1, 1)$, $(2, 5)$, $(-\frac{1}{2}, -7)$, $(0, -4)$?
$$\left(-\frac{1}{2}, -7\right), (0, -4)$$

2. Which of the following ordered pairs are solutions of $y \ge 6 - 2x$:
$(0, 0)$, $(1, 5)$, $(3, 6)$, $(\frac{1}{2}, 8)$?
$$(1, 5), (3, 6), \left(\frac{1}{2}, 8\right)$$

3. Which of the following ordered pairs are solutions of $4x + 2y \le 20$:
$(3, 0)$, $(0, 12)$, $(5, 5)$, $(2, -4)$?
$$(3, 0), (2, -4)$$

4. For the inequality $y < 5x - 2$, let $x = 0$. Solve for y.
$$y < -2$$

5. List 4 ordered pairs that are solutions to $y < 5x - 2$.
Possible answer: $(-1, -9), (1, -3), (0, -5), (3, 0)$

6. List 4 ordered pairs that are solutions of $3x + 2y > 4$.
Possible answer: $(0, 3), (2, 0), (-1, 4), (5, 1)$

56   **Holt Mathematics**

Problem Solving
Solving Two-Step Inequalities

Write the correct answer.

1. Grace earns $7 for each car she washes. She always saves $25 of her weekly earnings. This week, she wants to have at least $65 in spending money. At least how many cars must she wash?

at least 13 cars

2. Monty has saved $400 to spend on a video game player and games. The player he wants costs $275. The games each cost $39. At most, how many games can he buy along with the player?

at most 3 games

3. A video game club charges $8 per month as a membership fee, plus $2.75 for each game rental. Eugenie plans to join and rent no more than 5 games a month. What amount should she budget each month for video games?

at most $21.75

4. Cooper Middle School has a goal of collecting more than 1,000 cans of food in a food drive. So far, 375 cans have been collected. During the last 13 days of the drive, at least how many cans must be collected each day in order to meet the goal?

more than 48 cans

Choose the letter for the best answer.

5. In January 2005, it cost $0.37 to mail a letter weighing up to 1 ounce. Each additional ounce or part of an ounce cost $0.23. At most, what is the weight of a letter with $1.06 in postage?

A $w < 4$ oz
B $w < 3$ oz
C $w \le 4$ oz
D $w \le 3$ oz

6. Martin is planning a hedge along the back of his yard. The total length can be no more than 23 feet, and he will put a 4-foot-wide gate in the hedge. Each plant needs 2.5 feet of space to grow properly. How many plants should he buy?

F $n \le 7$
G $n < 7$
H $n \le 9$
J $n < 9$

7. The 12 members of the Middle School filmmaking club need to raise at least $1,400 to make a short film. They already have raised $650. How much more should each member raise on average?

A $x \ge \$62.50$
B $x \le \$62.50$
C $x < \$62.50$
D $x > \$62.50$

8. The rule of thumb in filmmaking is that you must shoot at least 3 minutes of film for every minute in a movie's "final cut." A 30-minute roll of film costs $250. How much will film cost for a 90-minute movie?

F $c \ge \$22,500$
G $c \le \$7,500$
H $c \le \$67,500$
J $c \ge \$2,250$

57   **Holt Mathematics**

Reading Strategies
Use a Flowchart

You can use a flowchart to solve a two-step inequality.

$$4x + 5 < 13$$

Step 1: Combine the terms without variables. Use the inverse operation.
$$4x + 5 - 5 < 13 - 5$$
$$4x < 8$$

↓

Step 2: Get the variable by itself. Use the inverse operation.
$$\frac{4x}{4} > \frac{8}{4}$$

↓

Step 3: Simplify.
$$x < 2$$

↓

Step 4: Graph.

−4 −3 −2 −1 0 1 2 3 4

Use the chart to answer the following questions.

1. What is the first step in the flowchart?

Combine terms without variables.

2. Why is 5 subtracted from both sides of the inequality?

to get 4x by itself on one side

3. What is the second step in the flowchart?

Use the inverse operation to get x by itself.

4. What operation is used to get x by itself?

division

58   **Holt Mathematics**

Puzzles, Twisters & Teasers

1-800-HELP!

Solve each inequality. Graph each solution set on the number line. Then use the letter next to your answer to solve the riddle.

1. $\dfrac{x}{5} - 6 < 19$

With what number did you start the graph?

125 = H

2. $\dfrac{y}{6} + 5 \le -13$

Is your circle open or closed?

closed = E

3. $-8x + 5 \le -51$

With what number did you start?

7 = I

4. $5y - 4 > -9$

Is your circle open or filled?

open = B

5. $\dfrac{x}{-3} + 8 > 11$

Is it possible for the solution to include −9?

no = N

6. $7y - 6 \ge 22$

Is it possible for the solution to include 4?

yes = L

Why were the alien's eyes so big?

He saw his P __H__ O __N__ E __E__
      125   no   ●

__B__ __I__ L __L__
O   7   yes

59

Holt Mathematics

Holt Mathematics